T0332472

MOLECULAR BASIS AND THERMODYNAMICS OF BIOELECTROGENESIS

TOPICS IN MOLECULAR ORGANIZATION AND ENGINEERING

Volume 5

Molecular Basis and Thermodynamics of Bioelectrogenesis

by

E. SCHOFFENIELS

and

D. MARGINEANU

University of Liège,
Laboratory of General and Comparative Biochemistry,
Liège, Belgium

KLUWER ACADEMIC PUBLISHERS

DORDRECHT / BOSTON / LONDON

Library of Congress Cataloging in Publication Data

Schoffeniels, Ernest.
Molecular basis and thermodynamics of bioelectrogenesis / by E. Schoffeniels and D. Màrgineanu.
p. cm. -- (Topics in molecular organization and engineering)
Includes bibliographical references and index.
ISBN 0-7923-0975-8 (HB : alk. paper)
1. Action potentials (Electrophysiology) 2. Bioenergetics.
3. Molecular biology. 4. Thermodynamics. I. Mărgineanu, D.-G.
II. Title. III. Series.
QP341.S36 1990
574.19'121--dc20 90-5344

ISBN 0-7923-0975-8

Published by Kluwer Academic Publishers,
P.O. Box 17, 3300 AA Dordrecht, The Netherlands.

Kluwer Academic Publishers incorporates
the publishing programmes of
D. Reidel, Martinus Nijhoff, Dr W. Junk and MTP Press.

Sold and distributed in the U.S.A. and Canada
by Kluwer Academic Publishers,
101 Philip Drive, Norwell, MA 02061, U.S.A.

In all other countries, sold and distributed
by Kluwer Academic Publishers Group,
P.O. Box 322, 3300 AH Dordrecht, The Netherlands.

Printed on acid-free paper

Printed in the Netherlands

TABLE OF CONTENTS

INTRODUCTION

Despite the fact that many years have elapsed since the first microcalorimetric measurements of an action potential were made, there is still among the research workers involved in the study of bioelectrogenesis a complete overlooking of the most fundamental principle governing any biological phenomenon at the molecular scale of dimension. This is surprising, the more so that the techniques of molecular biology are applied to characterize the proteins forming the ionic conducting sites in living membranes. For reasons that are still obscure to us the molecular aspects of bioelectrogenesis are completely out of the scope of the dynamic aspects of biochemistry.

Even if it is sometimes recognized that an action potential is a free energy-consuming, entropy-producing process, the next question that should reasonably arise is never taken into consideration. There is indeed a complete evasion of the problem of biochemical energy coupling thus reducing the bioelectrogenesis to only physical interactions of membrane proteins with the electric field: the inbuilt postulate is that no molecular transformations, in the chemical sense, could be involved.

It seems to us that such a position is untenable on purely thermodynamical ground: the cell membrane is a dissipative structure and bioelectrogenesis is no exception to it. The problem to be tackled is therefore that of chemo-electrical conversion. If the phenomenological aspect has attained today a high level of refinement and sophistication the main question to be answered is still that of the biochemical system providing the input of free energy required to fuel the impedance variation cycle at the origin of the action potential.

In the following pages we will bring forward the description of the action potential in the selfconsistent quantitative manner typical of physics, to come to the conclusion that bioelectricity is mostly an epiphenomenon which is caused by and in turn influences the metabolic dynamics of the cell.

This purely deterministic approach will also be applied to a brief analysis of the brain-mind problem since our wish is to contribute to bridge the gap still existing between on the one hand the biochemical and the biophysical approaches when dealing with bioelectrogenesis and on the other hand to show how cognitive science can benefit from the recent progresses in neurobiology.

Our warmest thanks are due to M. Rinné and R. Jacob from the Atelier Général de Métrologie Electrique et Electronique (AGMEE) of our University for the unfailing help they have provided us by composing the equations and setting up the whole text on the computer.

Thanks also to Mrs. M. Want who typed the manuscript and together with Mrs. I. Margineanu introduced the necessary corrections in the text.

CHAPTER I
THE DESCRIPTION IN PHYSICO-CHEMICAL TERMS OF NERVOUS SYSTEM PROPERTIES

Assuming that one should have to single out that philosophical, i.e. speculative, idea whose influence was most lasting and farest reaching, obviously there is no alternative to indicating the atomistic conception, put forward two and a half millenia ago and which, after more than two thousand years of dormancy, flourished becoming the very base of all natural sciences. At the dawn of this century, the atomistic view not only gained the respectability of an experimentally proved reality as concerns the structure of matter but it was extended to the other fundamental aspects of the physical world, the electric charge and the energy.

As concerns the biological sciences, following the virtual completion of the inventory of living forms on Earth, they became ever more concerned with the basic mechanisms of life, irrespective of the diversity of living beings. In the search of these mechanisms, biology has to rely on physico-chemical methods and thus, inevitably it became itself molecular. In departing from the vitalistic views, since the early nineteenth century it was recognized That living beings are subject to the laws of physics and chemistry. The contemporary omnipresence of physico-chemical explanations on the different phenomena from which life consists was in no way an unanimously acclaimed development. More than one oustanding traditionalistic naturalists vigorously refuted an approach viewed as too straight a jacket, unsuited for the multifarious biological diversity. If such an attitude is rather easily understandable on subjective grounds, the opposition to the inevitably materialistic molecular approaches in biology of some (even more outstanding) physicists seems less expected. Indeed, Niels Bohr extended his complementarity principle from the physical description of subatomic objects to claim that, in the study of living beings, sufficient degrees of liberty have to be left to the system for keeping it alive, that it will ever hidden the "ultimate secrecy of life". In his poetically inspired formulation, the existence of life has to be considered a fundamental "postulate". This position is unacceptable (and in fact ignored) by the stubborn practitioners of contemporary biology, whose individual contributions to science might lack grandeur, but whose collective achievements brought now the biology in the forefront of intellectual endeavour.

From physical point of view, the living organisms are extremely complex assemblies of molecules, able to extract free energy from the medium and to use it for their self-maintenance and conservative duplication. While the long term existence and evolution of the living forms rely on the genetic transmission of information, in the short run, each organism has to maintain itself against the entropic decay upon employing the energy taken from the outside for counteracting as far as possible the risks of external interactions and of internal degradation . Thus, on the one hand, all living organisms are dissipative structures and, on the other hand, they have to possess specialized systems to ensure a permanent gathering and processing of data needed to coordinate the function of all the parts of the system in a purposeful behaviour. These two rather general aspects have to be somehow settled, before touching the more specific ones and though it might seem a purely didactic job, we do it with the convinction that repeating (apparently) simple(1) fundamental things is to be preferred to their misuse.

I.1. Living systems as dissipative structures with local quasi equilibrium

Even from a superficial examination it is evident that biological systems reveal an organization: order is the rule in biology. The general tendency is to make order where disorder existed. This phenomenon is in apparent contradiction with the second principle which specifies that the natural evolution of a system proceeds by way of an increase of disorder. Is this to say that living systems violate the second law of thermodynamics (Schoffeniels, 1976; 1984)?

A long lasting but almost entirely sterile discussion revolved around the apparent discrepancy between the entropy increase due to the irreversibility of natural processes: $d/dt\ (S) > 0$ as it was postulated by R. Clausius, and the decrease in entropy of the living beings when they grow up and multiply. In fact there is no discrepancy at all, because the above formulation of the 2nd law holds only for isolated systems. All the living systems are open, their existence relying on the uninterupted exchanges of matter and energy with the environment. In this case, the time variation of entropy has no fixed sign, but it always will exceed the exchange with the outside (d_eS/dt):

♀

1 That nothing is too simple when dealing with "entropy" is proved by the continuous appearance of such rhetorically naive title as "What is the entropy?" (Prigogine, 1989).

$$\frac{dS}{dt} > \frac{d_eS}{dt}$$

because the system "produces" entropy: $d_iS/dt > 0$

The 2nd principle specifies that isolated systems evolve towards a state of equilibrium corresponding to a maximum of entropy, i.e. a maximum of disorder. For open systems, whose characteristic principle.is their existence far from an equilibrium state, the entropy is not maximum and there is a continuous flow of energy through them. In steady-state conditions, the energy flow from the source towards the sink evolves towards a constant value and the intensity factors of the different energy parameters (temperature, concentration, pressure, etc.) become equally constant with respect to time. In other systems undergoing an energy flow and which are far from equilibrium (without being in a stationary state), the exact significance of the intensity factors is not always clear.

Let us consider a stationary system formed by three compartments in series. The first compartment, 1, is the energy source and compartment 3 is the sink. The energy flows through the intermediary compartment 2. The second principle states that $dS_{13}+dS_2 > 0$, where dS_{13} is the entropy variation of the source and of the sink and dS_2 the variation of entropy of the intermediate system.

As dS_{13} is greater than 0 the sole limitation concerning dS_2 imposed by the second principle is that $-dS_2 < dS_{13}$.

Consequently the entropy of the system undergoing an energy flow can decrease. In other words the energy which crosses the system is utilized to produce the work necessary to maintain a state far from equilibrium. A biological system, statistically speaking, tends to attain the state which is the most probable, that is to say equilibrium (death). This situation is avoided, at least temporarily, but the utilization of energy to maintain the system in a less probable state, far from equilibrium. An isolated system is not able to perform work indefinitely. It is necessary that it be in communication with a source of energy and a sink. The work accomplished in the intermediate system is associated with an energy flow between the source and the sink, and, considering the system in its totality, the production of entropy is positive. When SchrÜdinger wrote that biological systems "feed on negative entropy", he meant to say therefore that their existence depends upon a continual increase in the entropy of the environment.

This concept has been proposed at a time when bioenergetics being still in its infancy, the reconciliation of the second principle and the biological fact was reached with difficulty. An organism feeds not on negative entropy, but on free energy which allows to maintain its entropy constant and possibly even to decrease. Since the whole process represents an apparent violation of the fundamental rule of the increase in entropy, the overall appearance is of a consumption of hypothetical negative entropy supplied by the environment. This is toying with words (always this fascination with words !) representing the state of ignorance within which the biologist and the physicist debated at the end of the Second World War. It is thanks to the work of Meyerhor and especially of Lipmann that the notion of coupling has been defined and that finally a little light has been thrown on the subject of transducers characteristic of the biological systems. If we do not know in their finest detail all the phenomena of coupling, we have nevertheless a very satisfactory view of the mechanism leading to the transformation of chemical energy into mechanical work (muscular contraction), light (bioluminiscence) and into osmotic work (active transport of ions), etc. Also, the transformation of radiant energy of the sun into sugars (photosynthesis) is, generally speaking, well understood (for a general account see Morowitz, 1977).

I.1.1. The creation of order by fluctuations

Thermodynamic analysis of the above considerations (Glansdorff and Prigogine, 1971) permits the demonstration that the energy flow through a system produces an organization of the system: there is creation of order. The simplest example illutrating this concept is furnished by Knudsen's membrane. Two compartments, 1 and 2, contain a perfect gas. The compartments are separated by a membrane possessing pores whose diameters are small when compared with the free mean path of the molecules. Compartment 1 is brought to a temperature T_1 different from that of T_2 in compartment 2. The membrane is adiabatic and the sole energy flow crossing this barrier is the movement of the gas molecules. The temperature gradient is localized at the level of the membrane. Under these conditions, the concentration of gas molecules in 1 will be different from that in 2. Another famous example is that of BÄnard's convection. At the beginning of this century, the French physicist BÄnard, observed at a critical temperature difference between the depth and the surface of a vessel, containing water, the formation of convection currents organized in the form of hexagonal network. The experiment is simple to perform: one places a vessel of sufficiently large surface on a warm plate taking care to avoid all other perturbations. One is dealing with the phenomenon of structuration corresponding to a high degree of molecular co-operation: the macroscopic convection currents are formed from a great number of molecules. To pass from the disordered state to the structured one, the system goes through an unstable state which is accompanied by an accumulation of energy. In order to maintain this state, energy must be continually fed in the form of a thermal gradient. Before the appearance of instability, the energy of the system is entirely defined by the disordered thermal agitation. After the appearance of instability at the critical temperature, the structures thus created are maintained only by continual exchanges with the external environment in conditions of nonequilibrium.

Prigogine gives the following physical interpretation of this phenomenon: in parallel with the increase in thermal gradient, small convection currents form which appear as fluctuations but which continually regress. The raising of the temperature approaching the critical gradient produces an amplification of some fluctuations which then give rise to the macroscopic current. This is the creation of order by fluctuation. The flow of energy between source and sink is associated with the formation of structures qualified as dissipative by Prigogine, and thus with the organization of the system.

The appearance of dissipative structures in non-linear systems far from equilibrium can be demonstrated theoretically and shown practically. The simple system just described presents interesting analogies with complex biological systems.

The virtue of its analysis lies in the demonstration that an energy flow is able to engender a molecular organization and that once order is attained, it is necessary continually to furnish the system with energy to maintain this state. In examples complicated progressively beyond that chosen above, it is possible to show that in simple molecular systems (H_2O, CO_2 and N_2 for example) undergoing an energy flow, one obtains through cyclical chain reactions compounds whose concentration is different from that theoretically predicted for stationary equilibrium conditions. The biological importance of this conclusion is evident since the concentration of intermediary compounds of a metabolic cycle are different from those presumed under the conditions of stationary equilibrium. Effectively this is observed in the case of glycolysis. The properties of the different enzymes constituting the sequence are sufficiently known to permit setting up effective mathematical models. Because the concentrations of chemical constituents participating in glycolytic reactions present oscillations, using the mathematical model, it has been possible to study the nature of these oscillations, which reveal a constant periodicity and amplitude. They originate beyond a zone of instability characteristic of a system in a stationary regime of flux far from equilibrium. This constitutes a temporal dissipative structure.

A dissipative structure is so organized that it increases its internal energy and dissipates more efficiently the flow of energy which traverses it. This proposal which implies a maximum accumulation of energy in the system finds its biological expression in the law of growth of an organism and in population dynamics.

Thus even in a simple physical system, the transfer of energy across the system causes an organization of this system. Once organized, in order to maintain itself in this state, the system must remain under the influence of an energy flow. A state far from thermodynamic equilibrium is thus installed which can undergo fluctuations and which is characterized by the formation of cycles.

Analysis of these systems according to the formalism of irreversible thermodynamics allows the conclusion, important for biologists, that the tendency towards organization is a very general property of certain classes of physical systems and is not specifically characteristic of living systems. Similarly it shows that biological systems imply instabilities which can be developed only far from thermodynamic equiliibrium.

I.1.2. Local quasi-equilibrium

Thus all living systems being spatio-temporal dissipative structures, goes along with the complementary idea that at much smaller space and time scales, there are quasi-equilibrium conditions. These make the local organization of the living matter at supramolecular level, to arise from the spontaneous tendency of the molecular components to attain that state which minimizes the free internal energy. The self organization of living matter on the basis of searching the thermodynamically most favourable configurations was amply discussed by Tanford (1978; 1988) and it can be best illustrated in the case of biological membranes.

That biological membranes are structures whose molecular components are essentially at equilibrium is suggested by their stability and confirmed by the spontaneous formation in proper conditions of artificial lipid aggregates, on/into which protein molecules adsorb, resulting in systems whose properties closely mimic those of the natural membranes.

The main thermodynamic aspect of their structure (see §II.2) is that it results from entropy-driven processes of self-assembly of amphiphillic molecules in an aqueous medium.

Almost a century ago, J. Traube observed that amphiphiles containing hydrocarbon chains lower the surface tension of aqueous solutions and concluded that these molecules preferentially go to the surface, the longer the chain, the higher being the surface concentration with respect to the bulk one. Later on, J.A.V. Butler and many others showed that the free energy of transfer of the non-polar molecules from an organic solvent to water is, predictably, positive, because of the preference of non polar molecules for non polar environments, but the enthalpy of transfer for many small hydrocarbons is, unpredictably, negative. In simple terms, this means that it is energetically advantageous for these non polar molecules to leave their medium and jump into water but, on entering water, there is a large drop in entropy which makes the transfer unfavorable. The entropy decrease shows that the non polar molecules have a greater orderliness when being in the aqueous environment. Accordingly, they tend to form aggregates, the simplest one being the monolayer which appears at the air-water interface.

The state of equilibrium in a complex chemical system can be defined precisely by Gibb's simple condition of the equality between the chemical potential of the components in all the phases or the places accessible to them. The (molal) chemical potential μ of one component (j) in a system is defined as the change in Gibbs free energy of that system, at constant temperature (T) and pressure (p), caused by an unitary variation in the number of moles (n_j) of that compound, i.e. it is the partial molal Gibbs free energy in isotherm-isobaric conditions:

$$\mu_{(j)} \equiv (\partial G/\partial n_j)_{T,p}$$

For an ideally behaving component of a solution:

$$\mu_{(j)} = \mu^\circ_{(j)} + RT\cdot \ln x_{(j)} \qquad (I.1)$$

where $\mu^\circ_{(j)}$ is the chemical potential in the standard state, in which j is alone in a pure phase, $\underline{R}$ = 8.31 J/mol, $\underline{R}$ is gas constant, $\underline{T}$ is the absolute temperature and $x_{(j)}$ is the mole fraction of j within the system.

The chemical potential of the hydrocarbon in the

et al., 1965) bilayers appear. As it was clearly described by Tanford (1978), the opposing thermodynamic preferences of the two ends of such a molecule are satisfied by self-association to form aggregates with the hydrocarbon chains in the middle, out of the contact with water.

The stability of these aggregates is often attributed to a "hydrophobic force". The term has to be employed simply as a label for the energetic and entropic factors which make $\delta\mu^{\circ}$ large and positive, thus preventing the hydrophobic and long-chain amphiphilic molecules from dispersing among the water molecules, but there is no special force, apart the usual intermolecular interactions (Hildebrand, 1979). In the case of alkyl chains, the forces are weak both in the hydrocarbon and in the aqueous media. The large negative contribution to $\delta\mu^{\circ}$ is mainly due to the fact that the alkyl chains reduce the number of hydrogen bonds between water molecules. Thus, the hydrophobic effect can be imagined as equivalent to a lateral pressure which prevents the hydrocarbon chains from coming in contact with water, squeezing them together. The same factors that make unfavorable the mixing of hydrocarbons and water, ensure the stability of phospholipid aggregates in water, and in particular the integrity of bilayers.

Along with the self assembly of lipids into bilayers, the incorporation of protein into membranes is also a spontaneous process, driven by a decrease in free energy as a result of the balance of two main contributions (Jahnig, 1983): a) the reduction of the mobility of water molecules surrounding the protein (i.e. the hydrophobic effect) which contributes a gain of - 147 kJ/mole of alpha-helix spanning the bilayer (Engelman and Steitz, 1981) and b) the partial immobilization of protein, which decreases this gain by about + 84 kJ/mol (Janin and Clothia, 1978), so that the binding energy for a hydrophobic alpha-helix is of the order of - 63kJ/mol. Such a low binding energy allows the regulation of protein incorporation between charged amino-acid residues and the membrane electric potential (Weinstein *et al.*, 1982). An essential point is that, though the proteins of membranes differ from the other water-soluble proteins in being partitioned between the lipid film (in which they are embedded) and the water (into which they protrude), the folding of their polypeptide chains is similar. The chains fold to alpha-helices and ß-sheets, whose outer surface is deprived of charges and have hydrophobic character. These portions of the protein traverse the hydrophobic interior of the bilayer, while the less regular regions connecting them have free hydrophilic groups exposed to the water molecules, with which they are hydrogen-bonded, or to the lipid polar heads.

The self-assembly of biomolecules into

an aqueous phase is positive and it linearly increases with the number n_c of carbon atoms in the hydrocarbon chain. The increment per each C atom is 3.7 kJ/mol for n-alkanes. In the case of aliphatic alcohols and acids, $\Delta\mu^\circ$ separates into a positive contribution from the hydrophobic moiety and a negative contribution from the hydrophilic head groups:

$\Delta\mu^\circ = - 3.49 + 3.44\ n_C$ (kJ/mol at 298 K) for aliphatic alcohols

$\Delta\mu^\circ = - 17.80 + 3.45\ n_C$ (kJ/mol at 298 K) for aliphatic acids

2) The unitary free energy of transfer, $\Delta\mu^\circ$, can be separated into enthalpic and entropic contributions:

$$\Delta\mu^\circ = \overline{\Delta G^\circ} = \overline{\Delta H^\circ} - \overline{T.\Delta S^\circ}$$

The unitary enthalpy difference, δH°, can be either measured calorimetrically or calculated from the dependence of μ° on temperature:

$$\Delta H^\circ = \left. \frac{\partial(\Delta\mu^\circ/T)}{\partial(1/T)} \right|_p$$

As previously mentioned, it is negative for small aliphatic hydrocarbons and for aliphatic alcohols.

3) The unitary entropy of transfer is negative for all hydrocarbons and for aliphatic alcohols, it being the main reason of the negligible solubility into water of hydrocarbons and of the low solubility of long-chain amphiphiles.

At extremely low concentrations of amphiphile, the translational entropy ensures dispersion, but, above very small threshold values, phospholipids will form aggregates to lower the total free energy. Whenever there is an aqueous medium a sufficiently high concentration of amphiphilic phospholipid molecules (containing a polar charged group at one end, attached to two hydrocarbon), organized supramolecular structures, in the form of either planar (Mueller *et al.*, 1962) or closed spherical (Bangham organic hydrocarbon phase is:

$$\mu_h = \mu^\circ_h + RT \cdot \ln x_h \qquad (I.2)$$

and in the aqueous phase, where its mole fraction is x_w, is:

$$\mu_w = \mu^\circ_w + RT \cdot \ln x_w \qquad (I.3)$$

The equilibrium condition states that the chemical potential of each definable component must have the same value in each phase or place accessible to it:

$$\mu_h = \mu_w$$

It follows:

$$\mu^\circ_h = \mu^\circ_w \equiv \delta\mu^\circ = - RT \cdot \ln(x_w / x_h) \qquad (I.4)$$

Equation (I.4) shows that the difference in standard chemical potential representing the unitary free energy difference ($\delta\mu^\circ$) is proportional the logarithm of the partition coefficient between the aqueous and the organic phases. In the special case where pure hydrocarbon, for which $x_h = 1$, is in equilibrium with the aqueous phase, eq. (I.4) become:

$$\Delta\mu^\circ = - RT \cdot \ln x_w \qquad (I.4')$$

indicating that the unitary free energy difference can be calculated from the mere measurement of the solubility of hydrocarbon in water.

Extensive analyses of data are given by Tanford (1973) and convenient summaries by Klein (1982) and Silver (1985). The following main conclusions will shed some light on the nature of hydrophobic effect (Margineanu, 1987):

1) The unitary free energy of transfer of hydrocarbons and long-chain amphiphiles from an organic to

supramolecular quasi-equilibrium structures is in no way restricted to the formation of membranes, but rather it is a general rule. It is quite obvious in the case of protein assemblies endowed with complex enzymatic and transport properties such as the (Na^{+}+ K^{+}) ATPases. A polypeptide chain spontaneously folds, getting secondary and tertiary structures, established by means of weak (electrostatic and H) bonds, then the globular protein thus formed associates with others to form a quaternary assembly with biological functions (Privalov and Gill, 1978). In a similar way, it was long ago shown by H. Fraenkel-Courat and R.C. Williams the spontaneous assembly *in vitro* of functional tobacco virus from two thousand identical viral proteins and one viral RNA molecule. More recently, the self -assembly of the bacterial flagella, of microtubules as well as of some extracellular collagen structures was described.

Summing up this point, it appears that the living beings are locally (i.e. at 10^{-9}m scale) structured so that to minimize the free energy, but always this extremum is conditioned by variable constraints which make the macroscopic system to be a dissipative structure supported by the surrounding universe from which a continuous input of free energy (or to which a transfer of entropy) must exist. Maintaining the large scale order within a huge assembly of molecules (of the order of 10^{38} in the human body) requires the operation of specialized communication networks.

I.2. Information flows in living systems

The coordination between the different subsystems of a complex system is made possible only if some kind of signals convey information about their actual states. Each organism is a subsystem in that milieu into which it can survive and the multicellular organisms (metazoa) have specialized subsystems which are the different cells, tissues, organs. Between all of these there are continuous flows of information, a task for which the nervous system is highly specialized, though it is not alone to perform it.

Though obvious as it might appear in view of its everyday use, the technical concept of information is only rather loosely defined as a measure of the indeterminancy of the actual state of a system which can be found with different probabilities in a multitude of states. Actually we dispose only of a rigorous mathematical definition for the amount of information, the generality of the concept and its anthropomorphic connotations preventing by now its complete definition.

Information being a physical action accompanying an effect on the receptor, the law of variation of information

(∂l) of a structure, during time ∂t, can be expressed as a function of the information flow due to exchanges of energy and matter with the environment $\partial_e l$ and to loss of information from irreversible processes within the same structure $\partial_i l$ as follows:

$$\partial l = \partial_e l + \partial_i l \qquad \text{with } \partial_i l \leq 0.$$

This equation is the extension to an open system of the law of variation of information as a function of time. It is homologous with the law of entropy variation of open systems as proposed by Prigogine.

We can establish a balance sheet for the information content of a biological system based on the exchanges which it effects in the course of its existence.

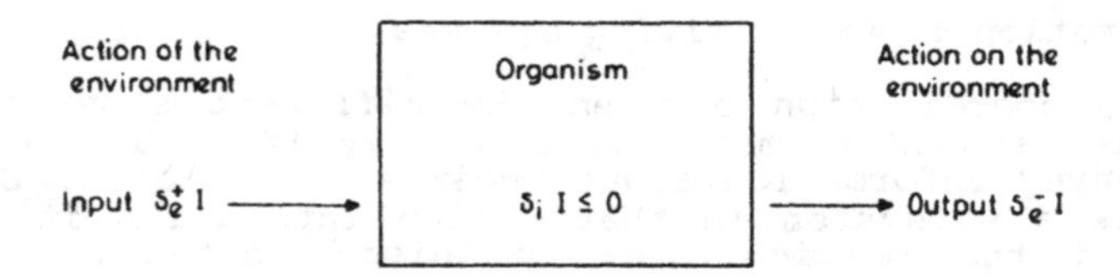

Figure I.1. Flow of information across an organism.

Figure I.1 represents the flow of information traversing an organism. The edge is the surface separating the environment and the organism. This envelope plays the role of selector of information both at the input (receptor) ($\partial^+ el$) and output (effector ($\partial^-{}_e l$)) levels.

Within the organism, degradation of information is represented by $\partial_i I$.

The selective nature of the envelope, with different receptors and multiple sources of information, is especially obvious in considering biological structures sensitive to physical (light, sound, etc.) and chemical (odour, taste, osmotic pressure, etc.) characteristics of the environment (Schoffeniels, 1976;1984).

The balance sheet is established in the following way. From its formation (fertilization) the organism possesses information Io representing an integration of information acquired in the course of evolution. As conceived in molecular biology, it is localized essentially in DNA. Information penetrating the organism thus permits: (1) maintenance of structure despite the irreversible degradation of information by irreversible phenomena. Such structure is essentially dissipative; (2) development of the structure (growth); (3) multiplication of the structure (reproduction); (4) a reaction on the environment.

In the case of the human and other animal species capable of training by apprenticeship, the intellectual information, the result of prolonged accumulation and interpretation of experience of preceding generations, represents a patrimony profitable to the various individuals of the community.

For the further discussions on the function of the nervous system, it will be necessary to precise the meaning of the different terms used when dealing with information processing.

By <u>signal</u>, it is meant any physical event expressing the state of a system and one which can be propagated through a particular medium. The variations in the physico-chemical parameters of the environment of a receptor cell (which we named stimuli) constitute signals which tell us something about the state of that environment. In their turn, the action potentials of the different excitable cells represent signals about the state of these cells. The system whose state changes is the <u>source</u>, generating signals. In the cases we mentioned above, the source is either the external medium or a receptor cell or even a sensory neuron.

The signal, for example a nerve impulse, is an elementary entity - by definition indivisible and repeatable. Usually no source generates single signals, but assemblies of signals constituting messages about the state of the source. For instance a train of nerve impulses occurring in a time interval at the initial segment of an axon represents a message about the state of the somato-dendritic membrane of that neuron.

According to their type of variation, the signals can be discrete (or discontinuous) such as the nerve impulses, or continuous like the potential variations of the receptor cells. The simplest communication system consists of a source of signals, a communication channel (the medium via which the signals are transmitted) and a receiver of signals. An appropriate scheme for the flow of sensory information is shown in Fig. I.2.

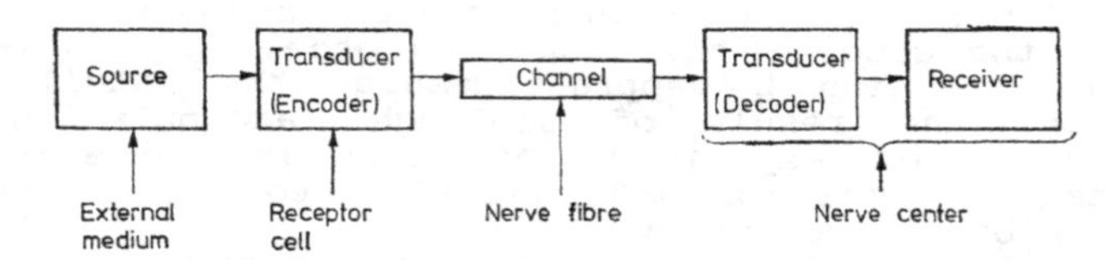

Figure I.2. Components of a generalized system for the transmission of information and their correspondance in the case of sensory information.

The messages (as assemblies of signals) carry information in the sense that, following their reception by the receiver, the previously existing, "ignorance" about the state of the source is removed. Therefore it is possible to speak about the information transmission through signals only when the state of the source is not known before the reception of the message. In other words, when a system generates signals whose time course is perfectly determined before their reception, it does not represent a source of information. Only random (or stochastic) signals can bear information.

I.2.1. Measure of the amount of information

In 1948 Ch. Shannon, by studying the process of communication via signals of any nature, defined the information carried by a signal. On reception of that signal a previous "doubt" about the state of the source is removed. In order to establish a quantitative measure of the degree of randomness of a signal, and thus of the information obtained by its reception, Shannon made the analogy with the degree of disorder of a statistical assembly of atoms and molecules as expressed by its thermodynamic entropy.

L. Boltzmann showed that for $\underline{N}$ identical particles (atoms or molecules) which can be distributed on $\underline{M}$ distinct energy levels, $\underline{N_i}$ on each level (i = 1, 2, ..., $\underline{M}$), the degree of indeterminacy in the real state of the particle assembly, and therefore the disorder degree, can be defined as:

$$S = k \sum_{j=1}^{M} \frac{N_i}{N} \ln\left(\frac{N_i}{N}\right) \qquad (I.6)$$

The quantity $\underline{S}$ so defined is the entropy of the assembly of particles; k is Boltzmann's constant, ($1.38.10^{-23}$ J/K); and the ratios N_i/N represent the probabilities pi of each energy level to be occupied, so that $p_i = N_i /N$. As:

$$\sum_{i=1}^{M} N_i = N$$, it obviously results that

$$\sum_{i=1}^{M} p: = 1.$$

It may be readily verified that entropy expresses the disorder in the assembly as:

- if the state of all particles is the same, the system being perfectly ordered: S = 0;
- if the system is perfectly disordered, each particle being on a separate energy level, M = N and $N_1 = N_2 ... = N_N = 1$, and the entropy has the maximum value: S= k.ln(N).

If a message is made up of N random signals (or symbols) of M different types, each of these types appearing in the message with the probability p_i (i = 1, 2 ... M), then the degree of indeterminacy of the message, i.e. its "entropy", will be:

$$H = -k \sum_{j=1}^{M} p_i \ln(p_i) \qquad (I.7)$$

The formula (I.7) is identical with (I.6), $\underline{k}$ being now another constant which depends on the chosen unit of measure. $\underline{H}$ expresses the amount of information carried by a signal in the respective message.

The most widely used unit for the amount of information is the <u>bit</u>: the information obtained on the reception of a signal when only 2 equally probable different signals can appear. In this case k = (1/ln2) and the formula of the amount of information expressed in bit is:

$$H = -k \sum_{j=1}^{M} p_i \log_2(p_i) \qquad (I.8)$$

If the signals are not discrete, but are grouped in a number of types $\underline{M}$, and they continuously vary as a function of a certain parameter x, then instead of the assembly of probabilities p_1, p_2 ... p_M of the appearance of each discrete signal, a continuous distribution with a density of probability $\underline{p(x)}$ exists so that:

$$\int_x p(x)dx = 1$$

In this case the amount of information is:

$$H = -\int_x p(x) \log_2 p(x)dx$$

Practically it is always necessary to pass from a continuous distribution to a discrete one, dividing the range of the parameter $\underline{x}$ into as large as possible a number of intervals and associating with each interval the probability that the continuous signal lies in that range.

I.2.2. The amount of information transmitted through trains of nerve impulses

On the basis of these concepts of the theory of information, the general scheme of receptor functioning

(Fig. I.3) can be interpreted as follows: the stimuli from the external medium constitute signals for the receptor cells, which in their turn generate signals with a continuous spectrum (the receptor potential and/or the generator potential). In the initial segment of the axon of the sensory neuron, a coding or modulation process occurs whereby a continuous variation of potential is converted into a discontinuous event - the generation of nerve impulses. The ensemble of these impulses is the carrier of the sensory information to the nerve centres where decoding processes and multiple processings take place (particularly at the level of synapses).

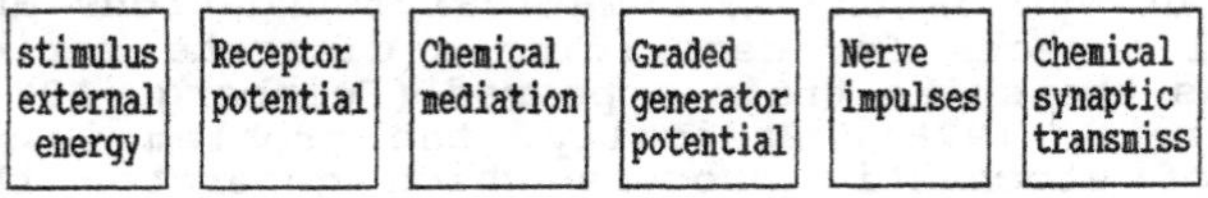

stimulus external energy	Receptor potential	Chemical mediation	Graded generator potential	Nerve impulses	Chemical synaptic transmiss

Figure I.3. Scheme of the events in sensory receptors functioning.

With respect to the temporal characteristics of a train of identical spikes, the main parameters which might play a role are shown in Fig. I.4:

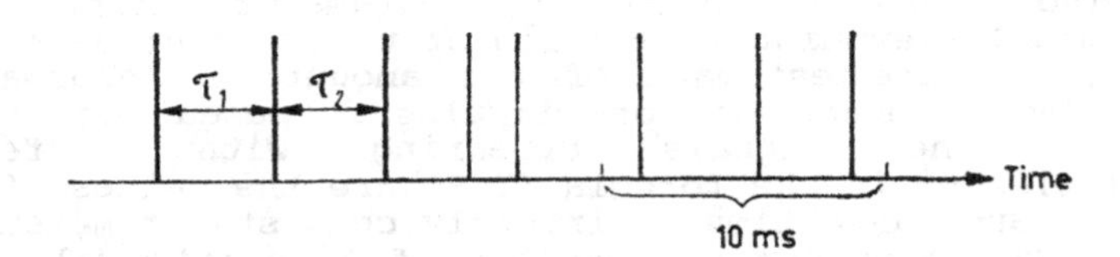

Figure I.4. Interspike intervals and frequency in a train of nerve impulses.

(1) The appearance time of each impulse or, what is equivalent, the inter-spike interval; this type of coding corresponds to what is known in technical language as position modulation.

(2) The frequency, i.e. the number of impulses in a given unit of time; this type of coding corresponds to the frequency modulation.

In a classical analysis of the maximum capacity of information transmission through the neuronal channel, Mac Kay and Mac Culloch have estimated this capacity on the basis of the values which the interval between impulses can reach. Several surveys of the mathematical foundations and of the technical aspects for describing a discrete series of events such as the spike train appeared (Feinberg, 1974; Landolt and Correia, 1978). Basically, the problem is to know the type of stochastic process which generates the train of impulses, that is the form of the functions characterizing the random variable (inter-spike intervals). There are several equivalent functions which can characterize the random inter-spike interval: (a) the probability density function p(x) which specifies the probability that an interval occurs between x and x + dx; (b) the cumulative distribution function:

$$P(x) = \int_0^x p(x)dx$$

is the probability that an interval will have a value not greater than X; (c) the probability R (X) that an interval will have a value greater than X is given by R(X) = 1-P(X). All these are estimated from the empirical interspike histogram established in a similar way as that for the amplitude of end plate potentials.

The most significant parameter from the physiological viewpoint in a train of nerve impulses is the frequency. In the estimate of the amount of information carried by a sequence of impulses, based on their frequency, the signals appearing with different probabilities pi in the formula (I.8) are the values of the frequency. Such an estimate virtually consists of measuring (for as large a number as possible of time intervals) the frequencies of impulses, of grouping them as a function of their values and of finding the probabilities of appearance of each value. Knowing these probabilities, the information transmitted via the neural channel can be calculated.

For instance, Griffith (1971) calculated that through each fibre in the optic nerve a maximum of 200 bit/s can be transmitted, the human brain receiving via the optic nerve an amount of information of the order of 108 bit/s and from all the receptors approximately 10^9 bit/s.

The practical usefulness of this kind of estimate is that more precise quantitative descriptions of the psychophysical processes can be obtained with a view to realize more integrated and tuned assemblies with man - technical systems for reception and processing of information.

1.3. Evolution of the ideas about bioelectrogenesis. A brief account

The generation of electricity by some very special organisms, the electric fishes, was certainly "felt" by many fishermen in the Mediteranean area, from where European culture originates, long before it became a subject of scientific inquiry related to the functioning of nervous system. When outlining here the history of the concepts about the production of electric signals by the excitable tissues, we shall not deal with the whole gamut of studies on the nervous system functioning, but mainly with the evolution of studies related to the generation and propagation of excitation in the nerve, its transmission between cells, and the response of the cells to external stimuli.

As in almost every aspect of physiology, understanding the nature of excitability has essentially depended upon the development of the physico-chemical methods and matured in the last three decades up to attaining the molecular level of explanations.

1.3.1. Prehistory of the study of excitability

Around 520 B.C., Alcmeon of Crotona described, on the basis of dissections, the optic nerve as a channel driving the sensation from eye to brain. Democritus (born about 460 B.C.) explained the function of sense organs on atomistic bases, stating that each surrounding object emits atoms which form an image of that object. It enters the sense organs, viewed as channels, and reaches the "soul".

The great physician of Rome, Galenus (131-201 A.D.), identified 7 (of the 12) pairs of cranial nerves and distinguished between the sensory and motor nerves. To explain the dominant role of the nervous system in the functioning of the whole body, Galenus considered that a part of the vital spirit resulting from mixing the blood with the lung air, reached the brain where it became animal spirit. This one, in fact the first ancestor of the nerve impulse, was thought as distributing through pores from cerebral ventriculi to all body compartments, especially the muscles and the peripheral sense organs.

The concept of "animal spirit", arising from that

of "vital soul" was used until the XVIIIth century to explain the nervous system functioning. René Descartes (1596-1656) stated that animal spirit proceeded from blood by filtration in the pineal gland and then flowed from cerebral ventriculi through nerves, which he imaginated as tubes with valves, allowing the regulation of flow. Despite the naive mechanicist character of Descartes' conceptions, his "animal spirit" is of material origin, it being a specific fluid which satisfies physical laws. Thus one can consider that Descartes anticipated the modern notion of nervous influx, though fanciful might seem today his descriptions. He also introduced the physiological concept of "reflex". The response to painful thermal stimuli is described as follows: heat stimulates the skin which excites the nerve fibres ending in the respective region. In their turn, these stimulate the brain zone they emerge from and open the cerebral pores. Through these pores the animal spirit penetrates, it reaching the muscle, where it elicits the contraction.

The Dutch naturalist Jan Swammerdam (1637-1680), showed that nerve stimulation leads to contraction even in isolated nerve-mucle preparations in which no animal spirit, produced by the brain, is involved. Nevertheless, the "animal spirit" concept was replaced by that of nervous influx of electrical origin almost two centuries later.

Within this "infancy stage" of the nerve impulse descriptions, Newton's (1642-1727) idea that the nerves are solid and transparent and that the excitation wave is propagated through them like a light beam through the "ether", might be worth reminding because, with a little bit of imagination, one can view it as anticipating data transmission in contemporary optic fibres technical networks.

I.3.2. Nerve impulse as bioelectric event

The idea of animal electricity was strongly advocated and experimentally supported by Luigi Galvani (1737-1798) whose paper *De viribus electricitatis in motu musculari* (1791) is considered as the coming into being of electrophysiology. Working with frog nerve-muscle preparations, he not only found that the externally applied atmospheric electrical stimuli can give rise to contraction in frog nerve-muscle preparations, but also that muscle contracts apparently without external stimulation when a circuit, made up of two different metals, is placed between it and its nerve. Galvani assumed that the nerve and the muscle would be charged with animal electricity which, when contacted by metals, produces the contraction. In 1774, H. Cavendish (1731-1810) had proved that the shock felt when touching the electric fishes (*Electrophorus, Torpedo, etc.*) was due to an electric current generated by them; therefore he put forward the idea that the animal organism produced electricity. He even built, by means of Leyden jars, a functional model of the torpedo, and improved this model until it produced the same kind of shock as the fish. This is the first known example of physical modelling in biological research.

Alessandro Volta (1745-1827) strongly contested the idea of bioelectrogenesis, he maintaining that in Galvani's experiments the frog muscle contraction was elicited by the electric current produced solely by metallic contact potentials. Later on, Alexander von Humboldt (1760-1859) showed that two different phenomena do indeed occur: an electric current generated by metallic contacts, but also, apart of this, a true intrinsic animal electricity production.

In 1843, Emil DuBois-Reymond (1818-1896), working with a high sensitivity magnetic needle galvanometer, discovered that between an injured fragment of nerve or muscle and an intact one there is a current which remains constant for a long time. Its presence shows that the injured segment is negatively charged with respect to the intact one. When the nerve or the muscle is electrically excited, the intensity of the "injury current" decreases, and there appear "action currents", whose sign shows the disappearance of the potential difference between the two segments.

In 1850, the great physiologist, physicist and mathematician Hermann von Helmholtz (1821-1894) realized the importance of measuring the conduction velocity within the nerve in an attempt to elucidate the unknown physico-chemical process underlying the propagation of the action potential. Working with a ballistic galvanometer, he succeded in measuring the very short time interval from the nerve stimulation to the muscle contraction and found that this varies proportionally to the nerve length between the stimulation electrode and the muscle. Thus, Helmholtz calculated a velocity of 25-30 m/s at room temperature, a value fully confirmed seventy-five years later, when cathode-ray oscillographs made possible the direct visualization of the latency interval between stimulus application and the appearance of excitation.

An important moment in the evolution of ideas about bioelectricity was the application (in 1902) by Julius Bernstein (1839-1917) of Nernst's theory of concentration piles for the description of biopotentials. He assumed the nerve membrane to be selectively permeable to K^+ in the resting state and explained the excitation through the sudden and reversible increase in membrane permeability to all ions. In spite of many shortcomings, Bernstein's basic idea that bioelectricity is the result of membrane phenomena, more precisely of ionic distribution between cytoplasm and extracellular medium, was of paramount importance, for it largely directed the subsequent development of neurobiophysics.

N. Rashevski (in 1933) and, independently, A.M. Monnier (in 1934) and A.V. Hill (in 1936) elaborated the theory "of the two factors". This theory assumed that in nerve fibres an excitatory substance and an inhibitory one coexist. Excitation was thought to take place when the concentration of the excitatory substance is higher than that of the inhibitory substance. An electrical stimulus would result in an alteration of the ratios of the concentration of both substances, the variations of the two concentrations being given by first order differential equations, with kinetic constants specific to each of them. The hypothesis of the two chemical factors, supposed to be involved in nerve excitation was elaborated under the influence of the famous Otto Loewi's experiments on the direct excitation and inhibition of heart by substances released through the stimulation of its afferent nerves. This hypothesis was later on abandoned and virtually forgotten, but within the scope of this monograph, we think it is worth recalling, as a remembrance of the old interest in the chemial processes involved in nerve transmission: the "chemical wave", in A.V. Hill's own terms.

The monograph published by A.M. Monnier in 1934 "L'excitation électrique des tissus. Essai d'interprétation physique" (Hermann, Paris), in which the experimental results on nerve excitability are described in detail and where electrical circuits equivalent to nerve are suggested, might be considered as a final point of the research on the nerve trunk as a whole, except for pharmacological purposes.

After 1934, F. Kato with his co-workers I. Tasaki and T. Takeuchi succeeded in isolating myelinated nerve fibres and demonstrated taht excitation is saltatorially driven from one Ranvier node to another. In 1936, J.Z. Young described the squid giant axon, which proved to be an excellent biological material for studying the nerve impulse. The very large diameter of this nerve fibre (± 500 μm in *Loligo* and up to 1500 μm in *Dosidicus gigas*) allowed the introduction of internal electrodes and the following up of the changes in axoplasm composition, brought about by impulse propagation; the electric potentials could then be correlated with the ionic flows.

With the availability of this remarkable biological material and the development of electronics, a more detailed picture of the processes taking place in the nerve during excitation evolved. In 1939, K.S. Cole and H.J. Curtis pointed out a decrease in the impedance of the giant axon during activity, which directly confirmed Bernstein's hypothesis that the axon membrane is muche more permeable during activity than in the resting state. In 1949, G.N. Ling and R.W. Gerard refined the technique of intracellular glass microelectrodes and, the same year, G. Marmont and K.S. Cole evolved the "voltage-clamp" technique. With this method, A.L. Hodgkin, A.F. Huxley and B. Katz analysed the dependence of ionic currents through the axon membrane on the transmembrane potential, then distinguished between the sodium component and the potassium one and described their specific kinetics. Their papers published in "Journal of Physiology" in 1952, represent the first attempt of describing a major biological phenomenon with the quantitative rigour, proper to physical chemistry. Within the frame of the ionic theory of excitation, the pharmacological dissection of the ionic currents with specific blockers, introduced after 1964 by T. Narahashi and developed by B. Hille and others, combined with tracer measurement of transmembrane ionic flows, allowed assessing the distribution of Na^+ and K^+ selective conducting sites and their respective conductances. In 1973, C.M. Armstrong and F. Bezanilla succeeded to measure the intramembrane charge displacement when conductances change, i.e. the gating currents postulated by Hodgkin and Huxley.

After the pioneering work of A. Verveen and H. Derksen, who reported (in 1965) spontaneous random fluctuations of the resting potential in frog Ranvier node, the study of membrane electric noise was extensively used as an indirect but sure possibility to distinguish within the ionic current, which crosses a certain membrane area, the contribution of individual sites. The significance of noise studies is truly major, as they represent, in the domain of membrane phenomena, the equivalent of brownian motion and indicate a real utilization of the concepts of molecular physics.

With the introduction around 1980, by E. Neher, B. Sackmann and colleagues of the impressive "patch-clamp" method, the electrophysiological studies became sufficiently refined to directly record currents flowing through a single ionic conducting site, with a time resolution of 10^{-5}s. Thus, it became available an unparalleled possibility to characterize the *in situ* behaviour of only one molecular quaternary aggregate of those membrane proteins which form the ionic sites - the transducers operating in cell membranes.

I.3.3. Studies of non-electric aspects of the nerve impulse

Quite understandably, the investigations on the metabolic connotations of nerve impulse generation and propagation are more recent than the bioelectric studies.

In 1913, S. Tashiro found an increased production

of carbon dioxide in a stimulated nerve and in 1917 he showed an increase in ammonia production following stimulation. W. Fenn (1927) established the increase in oxygen consumption of the stimulated nerve and the same year R.W. Gerard published a comprehensive study on nerve metabolism. These researches progressed within the broad context of investigations into cellular energy metabolism throughout biology. The proper determination of oxygen consumption in the active nerve was possible only after 1950, by using the polarographic type of "oxygen electrode", applied to the nerve as pioneered by F. Bring, Jr. and his group.

As far back as 1921, A.V. Hill began studies of nerve microcalorimetry which, with his co-workers, he pursued over four decades. Around 1970, B.C. Abbott, J.V. Howarth and J.M. Ritchie studied in detail the energetic aspect of excitation following up the thermal modifications occurring when one single impulse is propagated and succeeded in determining the oxygen consumption associated with the propagation of a single impulse.

In Ch. V, it will be shown at length that the experimental data on nerve mirocalorimetry stubbornly resisted the attempts to account them on sole physical grounds (derived from the ionic theory) and are a major reason for trying to complete this theory with biochemical aspects.

The study of the optical modifications accompanying the nerve impulse had been started as long ago as 1940 by F.O. Schmidt and O.H. Schmidt, but with little success, owing to the technical difficulties. Only when the electronic equipment became able to detect rapid variations in the intensity of light fluxes of about 10^{-6} of the reference value, i.e. around 1968, the groups of I. Tasaki, L.B. Cohen and G.N. Berestovskii showed evident changes of birefringence, light diffusion and fluorescence associated with the nerve impulse and interpreted them as proofs of some conformational alterations in the components of axon membrane. In spite of the hope attached to such studies, they provided rather little conclusive new data. The demonstration of the essential role of acetylcholine (ACh) as a chemical transmitter of excitation between cells, together with the fact that cholin-acetyl-transferase is quite frequent in nerves, led D. Nachmansohn to put forward in 1953 an enzymatic theory of excitation, according to which the excitation releases ACh from its bound form, which in turn triggers on increase in membrane ionic permeability. This theory was further reelaborated and several other biochemical schemes were proposed by one of us (E. Schoffeniels), all arising from the same basic idea, namely that it seems rather strange that nerve impulse would not involve any biochemical event directly associated with it. This topic is developed in Ch. IV. The correlation of the biochemical and electrophysiological studies of bioelectrogenesis was significantly facilitated by the use since 1957 of the single electroplax preparation (E. Schoffeniels and D. Nachmansohn), which provides an electrically excitable single cell quite rich in the molecular components involved in bioelectrogenesis.

I.3.4. Reception and transmision of excitation between cells

After 1920, E.D. Adrian and Y. Zotterman succeeded in recording nerve impulses on sensory fibres starting from the receptor cells; thus they were able to correlate the applied stimulus directly with the electrophysiological response. By such studies, subsequently made on different neuro-sensory systems, it was established directly that identical impulses in nerve fibres are transmitted no matter which stimulus elicited them, the specific interaction being only at the level of the receptor cells. The intimate nature of the primary processes, that is of the interaction between stimuli and the receptor cells is still rather obscure in many cases. A notable exception is that of the visual receptors, on the functioning of which a fairly detailed knowledge is available (Chabre and Deterre, 1989).

It appears that the basic sensory modes - sight and hearing - enjoyed the particular attention of the physiologists and their study developed in close connection with the corresponding fields of physics, namely optics and acoustics. The study of the reception of other stimuli was biophysically approached only much later, when the microelectrophysiological methods became sufficiently refined for allowing work with single receptor cells, isolated together with the afferent nerve fibre, and on which rigorously controlled stimuli were applied.

The way in which excitation is conceived to be transmitted within the nervous system and from it to the effector elements is directly conditioned by our concept of the microscopic architecture of the nervous system. Till the first decades of this century, many histologists considered that the whole nervous system represents a huge syncytial unit - the neurencytium. An alternative view is represented in the neuronal theory of Wilhelm His (1831-1904) and August Forel (1848-1931). The neuronal theory, essentially states that the nervous system is made up of neurons which are distinct cells, completely delineated by plasma membranes, each neuron being an individual functional unit with its own metabolic and functional dynamics, but lying in close interdependence with numerous other like units. Santiago Ramon y Cajal (1852-1934) showed that between neurons, and between neurons and the effector cells (which they innervate), though a very close contact exists, there are gaps termed "synapses" by C.S. Sherrington in 1897.

Structural continuity on the one hand and synaptic connections on the other created problems in terms of excitation transmission. It was necessary to postulate the existence of specific mechanisms. As early as 1857 Claude Bernard (1813-1878) showed that a strongly paralysing poison such as curare blocks the passage of nerve impulses through the junction of a nerve and mucle; it however does not affect the conduction of impulses in the nerve, nor the ability of muscle to contract in response to an electric stimulus directly applied on it. The immediate conclusion was that curare acts on some chemical factors operating at the neuromuscular junction. In 1874 DuBois-Reymond explicitly contrasted the chemical and the electrical transmission foreseeing the fact that nerves can communicate with the effector elements both directly by electrical signals and by the agency of substances released by nerve terminals.

The conception of neuromuscular transmision via chemical transmitters was strongly endorsed by the discovery that numerous substances, both natural and synthetic, are able to mimic the effect of nerve impulses, when either injected in blood, or locally applied at the level of junction. Thus, in 1909 J.N. Langley showed that nicotine stimulates the region of muscle end plates even when the degenerescence of nerve terminals occurred following the nerve section, and that the stimulating action of nicotine is blocked by curare in proportion to the relative concentrations of the two substances. He thus arrived at the concept of receptors which are located on the post-junctional membrane and which interact with the chemical transmitter.

In 1921, O. Loewi directly and convincingly demonstrated the chemical transmission from the parasympathetic vagus nerve (whose stimulation reduced the amplitude and frequency of heart beats) to the cardiac muscle, showing that in the perfusion liquid following nerve stimulation a substance appeared which was capable of directly inhibiting a second heart. The vagal substance proved to be acetylcholine, which H. Dale and his co-workers identified after 1936 as a chemical transmitter between motor nerves and striated muscles. In 1938 several researchers found that the first biopotential which can be recorded in the end plate zone, following nerve stimulation, is not the action potential of muscle fibres, but a local, non propagated, variation of potential induced by the chemical transmitter. In 1950 P. Fatt and B. Katz observed that in unstimulated end-plates it is possible to record the occurrence of spontaneous small local potentials whose amplitude is the multiple of a certain value.

The correlation of these observations with electron-microscopic pictures (E. De Robertis), showing the presence in the presynaptic axonal endings of some vesicles, led to the theory of quantal release of chemical transmitters. This theory explains the occurrence of the local depolarization of the end plate through the simultaneous releasing of a number of transmitter vesicles proportional to the frequency of nerve impulses. In the resting state spontaneous ruptures of vesicles take place, which elicits the miniatural potentials observed by Fatt and Katz. From 1965 onwards the studies of B. Katz and R. Miledi and of numerous other researchers aimed at the characterization of the processes occurring in the postsynaptic membrane under the influence of the chemical transmitter. J.C. Eccles described the ionic mechanisms which resulted in local and propagated potentials and subsequently the characterization of the ionic channels in the postsynaptic membrane was performed.

At the same time, the groups of E. De Robertis and J.P. Changeux studied the molecules playing the role of receptor on the postsynaptic membranes from the chemical and structural viewpoint. Pharmacological investigations revealed that different substances act as chemical transmitters in the central nervous system.

After 1960, strong evidence was produced concerning the existence in some specialized junctions of an electrical transmission between excitable cells. M.V.L. Bennett (1970) stated its physiological roles, namely the high velocity of transmission and the synchronous electrical activity of several cells.

All the processes involved in bioelectrogenesis, i.e. the excitability of the specialized receptor cells and nerve fibres, the propagation of the nerve impulse, and transmission of excitation between cells, chiefly arise from the properties of specialized cell membranes. The dynamic image of the structure of membranes, viewed after 1971 as "fluid mosaics" of lipids and proteins (in globular conformation) permits the interpretation of such processes within the framework of molecular transitions of membrane components. The studies related with the excitable cells were among those which contributed the most to an understanding of membrane property at the molecular level. At the same time, it becomes ever more apparent that the ability to generate voltage-dependent ionic currents which trigger or mediate various cell functions is shared by several cell types (endocrine, epithelial, immunocompetent), apart of the classical excitable cells, so that the importance of understanding the molecular bases of bioelectrogenesis is likely to get enhanced.

References

For the historical account in §I.3, the older original papers are not listed. As a general reference covering the developments up to 1980, Ch. I in the monograph by Vasilescu and Margineanu (1982) can be used.

Bangham, A.D., Standish, M.M. and Watkins, J.C. (1965) J. Mol. Biol. 13, 238-252.
Chabre, M. and Deterre, Ph. (1989) Europ. J. Biochem. 179, 255-266.
Changeux, J.P., Devillers-Thierry, A. and Chemouille, P. (1984) Science 225, 1335-1345.
Engelman, D.M. and Steitz, T.A. (1981) Cell 23, 411-422.
Feinberg, S.E. (1974) Biometrics 30, 399-427.
Glansdorff, P. and Prigogine, I. (1971) Thermodynamic Theory of Structure, Stability and Fluctuations. Wiley Interscience, London.
Griffith, J.S. (1971) Mathematical Neurobiology (Ch. 6) Academic Press, New York.
Jèhnig, F. (1983) Proc. Natl. Acad. Sci. USA 80, 3691-3695.
Janin, R.A. and Clothia, C. (1978) Biochemistry 17, 2943-2948.
Klein, R.A. (1982) Q. Rev. Biophys. 15, 667-757.
Landolt, J.P., Correia, M.J. (1978) IEEE Trans. BME 25, 1-9.
Lester, H.A. (1988) Science 241, 1057-1063.
Margineanu, D.G. (1987) Arch. Internat. Physiol. Biochim. 95, 381-422.
Morowitz, H.J. (1978) Foundations of Bioenergetics, Academic Press, New York.
Mueller, P. Rudin, D.O., Tien, H.T. and Wescott, W.C. (1962) Nature 194, 979-980.
Numa, S. (1986) Biochem. Soc. Symp. 52, 119-143.
Prigogine, I. (1989) Naturwiss. 76, 1-8.
Privalov, P.L. and Gill, S.J. (1988) Adv. Prot. Chem. 39, 191-234.
Schoffeniels, E. (1976) Anti Chance, Pergamon Press, Oxford.
Schoffeniels, E. (1984) L'Anti-Hasard (Japanese translation), Misuzy ShobÖ, Tokyo.
Silver, B.L. (1985) The physical Chemistry of Membranes (p. 111), Allen and Unwin, Boston.
Tanford, C. (1973) The Hydrophobic Effect. Wiley, New York.
Tanford, C. (1978) Science 200, 1012-1018.
Tanford, C. (1987) Biochem. Soc. Trans. 15, S1-S7.

Toyoshima, C. and Unwin, N. (1988) Nature 336, 247-250.
Unwin, P.N.T. and Zampighi, G. (1980) Nature 283, 545-549.
Varner, J.E. (Ed.) (1988) Self-Assembling Architectures, 276 p. Alan R. Liss, Inc. New York.
Vasilescu, V. and Margineanu, D.G. (1982) Introduction to Neurobiophysics, Abacus Press, Tunbridge Wells, Kent.
Weinstein, J.V., Blumenthal, R., Van Renswonde, J., Kempf, C. and Klausner, R.D. (1982) J. Membrane Biol; 66, 203-212.

CHAPTER II
CELL MEMBRANES AND BIOELECTROGENESIS

The most basic property of all living organisms to process matter - for extracting from their milieu the free energy, necessary for their maintenance and growth-, imposes the existence of well-defined domains of the Universe, in which proper concentrations of reactants and products are maintained against the entropic tendency of dilution. This fundamental requirement was recognized more than one and a half century ago by Theodore Schwann and repeatedly stressed since that. It is confirmed by the fact that all the cells, including the most primitive prokaryotes, are surrounded by a continuous and flexible structure - the plasma membrane - which appears as a distinctive feature of living system.

II.1. The ubiquitous cellular component

Plasma membrane are extremely thin (of only 40 to 100 Å, so that they fall much below the resolution limit of the light microscopy), but tenaciously stable structures, which define the space into which the metabolism occurs.

As their thickness is by several orders of magnitude smaller than the lateral dimensions, plasma membranes appear as truly <u>two-dimensional</u> objects. Their omnipresence, together with the fact that their supramolecular architecture is strikingly similar in all cells, suggest that the appearance of such structures was an early event in the evolution leading to life. It might be viewed as the transition step between the unidimensional biopolymers and the three-dimensional structure of the whole cell.

Always the biomemembranes appear as <u>closed surfaces</u>, forming enclosures which establish a deep compartmentalization of the cell even inside the cytoplasm. Thus, apart of preserving the distinct composition of the cell, the membranes act as inter- and intracellular boundaries which make possible the simultaneous and non-random, but ordered occurrence of a large number of reactions.

Along with preventing the dilution of cell components and of regulating the flows of matter, the membranes are the specific site of many energy conversions. In fact, with the only exception of the chemo-mechanical conversion, which is performed in filamentous contractile structures, all the other types of biological energy transductions occur in membranes. These provide the highly organized molecular frame into which the non-radiative and only moderately dissipative transfer of the electronic excitations is readily possible.

The fact that plasma membrane is the outermost cellular component implies that it should be the site of recognition of external signals, be them molecules or fields of various forces, thus being involved in data processing. Depending on their specialization, the plasma membranes possess particular receptor structures for detecting the signals and transduction and amplification equipment for the response. In spite of the diversity of stimuli, the detection cannot rely but on reversible conformational changes of some membrane components endowed with plasticity of shape. The transduction and amplification are achieved because the stimuli trigger the release of a store of metabolic energy, either by changing the activity of some membrane bound enzymes, the adenylate cyclase being a most significant example (Schram and Selinger, 1984), or by abruptly changing the electrical conductance of the membrane.

Because it is the external coat of the cell, plasma membrane must be stable (in order to keep a near constant internal volume and to prevent the free mixing of the two aqueous media it separates) but at the same time it must tolerate without rupturing sometimes dramatic shape changes (as for an erythrocyte passing through capillaries) and to allow the insertion of newly synthesized and the recycling of altered components. To cope with the apparently conflicting requirements of stability without rigidity, the membranes must be ensembles of molecules in (local) equilibrium, i.e. at a minimum value of free energy, but without strong attractive intermolecular forces, as it was discussed in § I.1.

All living cells are virtually isothermal and isobaric systems as no noticeable differences in temperature and hydrostatic pressure appear, even in ectothermic organisms, on a timescale comparable with that of the metabolic processes. Accordingly, the membranes themselves can be treated as open, isothermal and isobaric systems, though the very definition of such macroscopic parameters as temperature and pressure might rise some conceptual problems in the case of molecular size systems (Margineanu, 1987).

Their role as selective barriers between two aqueous media dictates the gross composition of cell membranes. To make these media immiscible for preserving the molecular frame of life, membranes have to be mostly hydrophobic and thus of lipidic nature. As lipid molecules with to opposite polar heads, suited for the contact with the two aqueous compartments, are but exceptions, the membranes must in general be formed by two lipid monolayers juxtaposed. However, in extremely thermophilic bacteria, membrane lipids consist in two hydrocarbon chains covalently bound at both ends to two polar heads. In order to allow the exchange of hydrophilic metabolites and wastes of the cell with the external medium, the membrane must possesses hydrophilic zones acting as channels and carriers. These zones cannot be but high molecular weight proteins, because only such macromolecules can constitute hydrophilic structures or undergo the rapid reversible conformational changes involved in carrying a substrate.

Obviously, any attempt to derive from basic requirements the characteristics of biological membranes is a typical *a posteriori* reasoning, from which is not to be expected the discovery of new facts. However, the effort to find the "raison d'être", that is to understand "why Nature made the things in a certain way", is of chief organizing value and lies at the core of scientific explanations.

II.2. Membrane molecular components and their dynamics

The chemical species occurring in biological membranes are mainly high molecular weight compound: proteins, lipids and carbohydrates, but also water and small inorganic ions.

In recent years, a vast literature accumulated, suggesting that water and inorganic ions play an important role in the maintenance of structural integrity and properties of biomembranes (Franks and Finney, 1981; Pullman *et al.*, 1985). Detailed investigations of membrane changes during gradual drying were performed, in connection with the remarkable ability of a number of organisms to survive nearly complete dehydration (Crowe and Crowe, 1982a,b). In spite of these, little emphasis has by now been placed on water as a true membrane component, it rather appearing - due to its huge proportion within and around the cells - as the molecular milieu in which the organic components are organized into supramolecular structures.

The carbohydrate usually accounts for (1-10 %) of the total dry weight of membranes and is covalently bound to either lipids or proteins, forming the glycolipids and glycoproteins. These have their oligosaccharide side chains only on the outer face of the membrane (Lennarz, 1980; Grant, 1984) and thus are involved in cell-cell interactions.

In practically all membranes, around 95 % of the bulk is represented by the lipids and proteins, whose proportions vary greatly. The protein content ranges from only 22 % in the case of central nervous system myelin, up to 70 % in Ehrlich ascites cell membranes, or to even more than 75 % in the inner mitochondrial membrane (details on membrane composition were repeatedly reviewed: Kotyk and

Janacek, 1977; Benga and Holmes, 1984). As a general rule, in most cell membranes, the proteins are the larger fraction by weight. The actual chemical structure of the membrane components, appearing in many recent textbooks, such as by Sandermann (1983), Alberts *et al.* (1983), is of lesser significance for the purpose of this book, what matters being that, in spite of their chemical differences, all lipids are hydrophobic, oil-soluble compounds. As for the charge on their molecules, some membrane lipids are uncharged (e.g. cholesterol, triglycerides), others are zwitterionic (e.g. phosphatidylethanolamine, phosphatidylcholine). Also, some are weak acids (phosphatidylserine), while others are strong acids (phosphatidic-, sulpho acids). With respect to the dipol-moment, some of the major membrane lipids, such as the phospholipids and the glycolipids, have a polar head and two non-polar acyl chains, but some other equally frequent membrane lipids, such as the cholesterol, are non-polar.

Since the masterly elegant experiments of Gorter and Grendel (1925), the membrane lipids are viewed as forming a bimolecular leaflet with the polar heads pointing outside to the aqueous cytoplasmic and external solutions, and the non-polar tails of the two monolayers facing each other. The currently popular "fluid mosaic model" of membrane structure (Singer and Nicolson, 1972) envisages the lipid bilayer as a two-dimensional fluid in which float the membrane proteins. Some of these, the peripheral or extrinsic, are only loosely attached on the bilayer, from which they can be removed by mild treatments with saline solutions and they solubilize free of lipids. Other membrane proteins traverse the lipid bilayer (integral proteins) and require hydrophobic bond-breaking agents such as the detergents to extract them out of the membrane and upon solubilization remain associated with lipids (Singer, 1974).

The integral membrane proteins are generally hydrophobic and thus rather difficult to handle. Upon compiling the primary structure of several tens integral proteins, ClÄment (1983) distinguished two modes of insertion in membrane: i) the integration by a single peptide which is longer and more hydrophobic than uncharged peptides of soluble proteins and is limited by several positive charges on the C-terminal side; ii) integration by a complex folding inside the phospholipid bilayer. In this latter case, the knowledge of the amino-acid sequence is not sufficient to predict the manner in which the protein interacts with the membrane.

Parts of transmembrane proteins span the hydrophobic regions of the bilayer as alpha-helically coiled polypeptides. The proteins that span different types of surface membrane belong to 2 distinct classes:

1) One type has a single membrane-spanning segment, in alpha-helical conformation, made of 22-23 uncharged amino acids, just long enough to cross the nonpolar region of the membrane. To this class belong the membrane bound immunoglobulins and a variety of receptor molecules, including those for insulin (Ullrich *et al.*, 1985).

2) Another type has multiple polypeptide chains crossing the lipid bilayer. The membrane-spanning segments of this class of molecules has charged amino-acids within the hydrophobic regions. These proteins are involved in different transport functions and form "channels" across the bilayer. The charged amino acids face the "pore" of the channel, which is hydrophilic.

Some integral proteins consist in a single unit (e.g. the bacteriorhodopsin), but in principle, several globular molecules may associate within the plane of the membrane to form composite organisation such as dimeric (cytochrome c oxidase) or hexameric structures (Ach receptor) (Amos *et al.*, 1982). Recently, schematic models for the arrangement and folding of the polypeptide chain in the bilayer were proposed for the electrophysiologically very important acetylcholine receptor (Guy, 1984), and the voltage-sensitive sodium channel (Greenblatt *et al.*, 1985), on the basis of a variety of computer-aided analyses.

It has been realized for some time that models that only deal with integral membrane proteins do not explain how cell surface properties are regulated, nor how membranes can undergo dynamic changes in their shape, deformability and stability. It is now clear that a complex set of proteins is attached to the undersurfaces of cell membranes and is responsible for stabilizing membranes and regulating the topography and mobility of the different transmembrane proteins. This "membrane skeleton" is composed of filaments of actin, actin-binding proteins such as spectrin and a set of connecting proteins that link the stabilizing infrastructure to the overlying membrane (Marchesi, 1985).

The fact that no strong attractive force among its components exists (ñI.1) makes the membrane not only deformable, but also fluid. Striking evidence for redistribution of membrane components initially came from the developmental biology of nerve and muscle membranes (Hubbell and McConnell, 1968 and 1969) and from the measurements of intermixing of cell surface antigens after fusion of mouse and human cells (Frye and Edidin, 1970). The spreading of surface antigens in newly fused heterokaryons provides a spectacular direct proof of the ease with which membrane proteins diffuse in the plane of the membrane and is an indirect demonstration of the fluidity of the lipid bilayer.

The relative mobility of lipids and proteins in

cell membranes was intensively investigated and repeatedly reviewed in the last decade (Shinitzky and Henkart, 1979; Edidin, 1981; Quinn, 1981; Axelrod, 1983; McCloskey and Poo, 1984) because it is essential for normal life processes (e.g. cell growth and regeneration) and its modifications are involved in a large spectrum of pathological conditions (Lenaz, 1984) including malignancy (Shinitzky, 1984). For the following discussion of the thermodynamics of bioelectrogenesis, it is to be noted that intramembrane molecular motions are a prerequisite for the operation of both conducting sites and pumps.

Both the integral and the peripheral proteins are kept in/on the membrane by weak forces, the hydrophobic and the ionic interactions, respectively. The energy of coulombic ionic interactions is around 17 kJ/mol for unit charges in water, and that of the hydrophobic interactions around 25 kJ/mol. Both these, and the Van der Waals attractive forces between nonpolar residues, whose energies also do not exceed 15 kJ/mol, are weak forces as compared with the covalent bonds, whose energies are around 375 kJ/mol. The fact that membrane component are not tightly held together, but allow many relative displacements, is a chief thermodynamic feature of biomembranes.

As a rule, the proteins are randomly distributed within the membrane plane, but in some particular cases they exhibit quasicrystalline order, for instance in the purple membrane of *H. halobium* (Henderson and Unwin, 1975) or in the photosynthetic bacterium *Rhodopseudomonoas viridis* (Welte and Kreutz, 1982). Sometimes, even the lipid hydrocarbon chains have a considerable degree of order (Jackson and Sturtevant, 1978). If the function of the cell imposes the existence of specialized membrane areas, as it is the case of neurons, the proteins are not at all randomly distributed within the plane of the membrane.

Independently of such differences, some membranes might be reasonably considered a continuum material in the two dimensions of the surface, this being the core assumption of the thermo-mechanico-chemical studies dealing with membrane elasticity and viscosity (Evans and Hochmuth, 1978; Evans and Shalak, 1980). In contradistinction to the (relative) isotropy in the surface plane, the anisotropy along the thickness is a general rule which cannot be neglected. Most lipid species are found in both monolayers of the membrane, but at different concentrations (Op den Kamp, 1979). For proteins, the asymmetry on the two sides is absolute (Chapman and Hayward, 1985) and refers not only to the existence of quite distinct peripheral proteins on the inner and outer faces, but also to the fact that the trans-bilayer integral proteins have structurally and functionally different parts inside and outside. On these differences relies the vectorial, space oriented characteristic of the flows of matter and of signal transductions in membranes.

The biosynthesis of phospholipids occurs in discrete subcellular compartments on the cytoplasmic side of the endoplasmic reticulum bilayer, from where they are transported and then they assemble into asymmetric bilayers (Bishop and Bell, 1988). As concerns the synthesis of membrane proteins, it occurs quite often rather far from their final location. Thus, in a neuron all protein synthesis, including that of the membrane ionic conducting sites, must take place in the cell body, because only there are ribosomes, endoplasmic reticulum and Golgi apparatus. The proteins move along the axon in the form of intracellular membrane vesicles, at a speed of 40 cm/day (Graftein and Forman, 1980), these vesicles fusing with the plasma membrane to which they transfer the proteins.

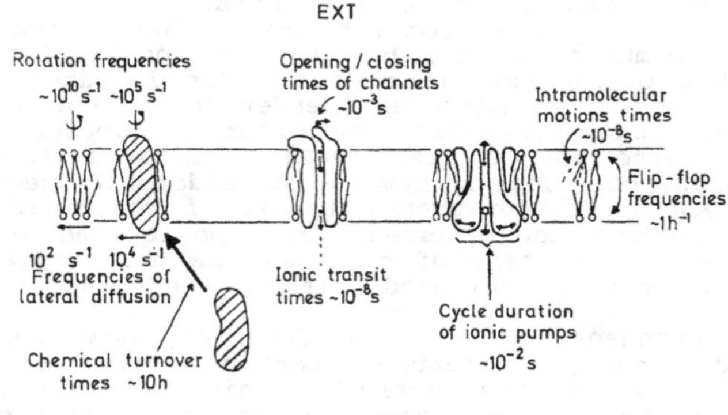

Figure II.1. Time characteristics of the molecular events within the biological membranes.

The vaguely defined term "membrane processes" covers several kinds of phenomena, which proceed with quite different time characteristics (Fig. II.1). The knowledge of the timescale imposes definitive restrictions on the approaches suited in each case, because some events are so slow as they might be treated as quasi-static, equilibrium processes, while others are so rapid as to preclude the use of thermodynamic concepts.

The membrane molecular components are subject to a chemical dynamics, they being continuously destroyed and replaced, even in mature fully differentiated cells, as shown by the steady decline in radioactivity after pulse-labelling *in vivo*. The biological concept of <u>turnover</u>, as it has been epitomized in Schoenheimer's (1942) classical book "The dynamic state of body constituents" is reminiscent of Heraclitus' (500 B.C.) idea of the continual flux amidst apparent constancy. The metabolic turnover rates are measured by administering to the organism a radioactive precursor of each kind of membrane component and determining the half life $T_{1/2}$, i.e. the time during which half of the radioactivity is lost (Schimke, 1975). Because $T_{1/2}$ of the different membrane components varies widely (see for instance, Finean *et al.*, 1984), it is easy to conclude that the membranes are not usually replaced as whole domains. For plasma membrane and endoplasmic reticulum proteins, $T_{1/2}$ values are around 50h, but some enzymatically active proteins such as the $(Na^+ + K^+)$ ATPase may be renewed every few hours (Cook *et al.*, 1976), while others last several weeks (e.g. NAD glycohydrolase). The turnover rates of membrane lipids are somewhat less different, the reported values of $T_{1/2}$ ranging from 14 to 140 h (Siekevitz, 1975). Along with other membrane proteins, the ionic sites are continuously internalized and degraded in lysosomes, where they are hydrolyzed to amino-acids. This was extensively demonstrated in the case of synaptic receptors for acetylcholine in muscle cells in culture, which have a mean lifetime around 30 h (Pumplin and Fambrough, 1982). The Ca conducting sites which operate in the membrane of *Paramecium* were estimated to have a lifetime ranging between 11 and 15 days (Schein, 1976).

The physical dynamics of membrane components consists in (at least) three types of intramembrane movements of the whole molecules (reviewed by Edidin, 1981; Kotyk, 1985), apart of the intramolecular motions (Thompson and Huang, 1985). The axial rotation of the lipid molecules within the monolayer has rotational frequencies of 10^9 - $10^{10} s^{-1}$, while that of proteins is characterized by significantly lower values: 10^4 - $10^6 s^{-1}$. The lateral diffusion along the plane of the membrane has diffusion coefficients of $(10^{-13}$ - $10^{-11})$ $m^2 s^{-1}$ corresponding to about 10^7 jumping events per second in the case of lipids, and $(10^{-16}$ - $10^{-13})$ $m^2 s^{-1}$, corresponding to 10^4 jumps/second in the case of proteins. In most cases, proteins were found to diffuse within plasma membranes at considerably lower rates than those seen in artificial membranes (Jacobson *et al.*, 1987) and it was established that increased membrane protein concentration decreases their lateral diffusion coefficients. The flip-flop movements from one to the other monolayer of a membrane are extremely unlikely, having frequencies of only (0.1-10)h-1 for lipids, and being virtually non-existent for proteins. Transmembrane phospholipid flip-flop is energetically unfavorable and does not occur spontaneously because of the high energy barrier faced by the polar head when crossing the hydrophobic bilayer.

Since 1977, it was shown that the sites for sodium are non-uniformly distributed over the surface of nerve cells, they being highly abundant in the Ranvier node and virtually nonexistent under the myelin sheath (Ritchie and Rogart, 1977). This clearly means that the mobility of these protein aggregates within the two-dimensional membrane fluid is much restricted to distinct areas, this being a necessary condition for the maintenance of the polarity of the cell. Recently, Angelides *et al.* (1988) employed fluorescent photobleaching recovery and fluorescent labelled neurotoxins, specific for sodium conducting sites, and found that these are free to diffuse within the cell body (of cultured neurons) with lateral diffusion coefficients of the order of 10^{-9} cm^2/s. But the ionic sites are somehow prevented from diffusing between cell body and the axon hillock, where the diffusion is significantly slower ($D \approx 10^{-10} - 10^{-11}$ cm^2/s). However, no regionalization or differential mobilities were observed for both lipid and glycoprotein diffusion. The segregation of sodium conducting sites into local compartments is probably ensured by the attachments with the cytoskeleton.

The intramolecular motions of membrane components occur on a timescale ranging from 10^{-5} to 10^{-9}s. The significant interaction which exists between the membrne lipid and protein was demonstrated by observing (with NMR relaxation techniques) the lipid over this range of timescales (Cornell *et al.*, 1983). These authors found that the timescale of the molecular motions within the intact natural membrane is strongly depressed to lower frequencies when compared to either a pure synthetic lipid, or to a lipid extract of the membrane. The 10^{-9}s motion of the lipid chain is depleted and that of 10^{-5}s is far more intense. (The low frequency motion involves collective movements of large segments of the hydrocarbon chain, which were suggested as a possible source of lipid-mediated protein-protein interaction).

The transmembrane movement of micromolecular solutes occurs in time intervals ranging from 10^{-8} to 10^{-12}s, depending on the mechanism of permeation. Thus, through the sodium and the potassium selective sites, opened by an electrical stimulus in the excitable axonal membrane (§ II.3) there is an inflow of 10^7 -10^8 ions/second (Hille, 1984). For a single file diffusion, this implies a time lag of ≈ 10^{-8}s needed by an ion to cross the axolemma. Rather similar data are for the <u>ionic channels</u> created in lipid bilayers by the incorporation of gramicidin A (Finkelstein and Andersen, 1981).

The ionic conducting sites present in the axonal and the postsynaptic membranes undergo conformational changes which shift their conductances between at least two (but quite often more) conducting states: a low conductance state (LCS; closed) and a high conductance (HCS; open) one. The conformational changes occur stochastically on a timescale of 10^{-5} to 10^{-2}s, the mean lifetime of the open state being most often around 10^{-3}s (Neher and Stevens, 1977; Auerbach and Sachs, 1984). The vibration of peptide backbone of the intrinsic protein are much too fast to be relevant for the operation of the conducting site, which open and close due to major readjustement of the tertiary and quaternary structure. The nature of gating mechanism is a central point of our concern around which much of our discussion will revolve.

If a solute molecule is transported through the membrane by a carrier, such as the valinomycin transporting K^+, the passage is considerably slower than in the case of conducting sites, of less than 10^4 ions/second (Benz *et al.*, 1973).

The anti-entropic transfer of ions against their concentration gradients, directly coupled with the exergonic ATP hydrolysis, occurs at an even slower rate. Thus, the maximal turnover of the plasma membrane sodium pumps is around $150s^{-1}$ (Karlish and Pick, 1981).

In conclusion, one can distinguish the following main groups of (intra) membrane events, with quite different time characteristics: i) the chemical turnover of membrane components; the timescale of these events extends up to several days, they being sufficiently slow as compared with the intramembrane motions to treat the system as being in quasi-equilibrium and to use the concepts of classical (chemical) thermodynamics ii) the transmembrane mass and charge transfer and the associated energy conversions; the time intervals in which one molecule (ion) traverses the membrane range is about the same timescale as the intramembrane molecular motions, so that for the description of the associated macroscopic flows it is necessary to apply nonequilibrium thermodynamics iii) the unitary molecular events, such as the electronic excitations as a result of light quanta absorption or of resonant transfer, and the intramolecular conformational changes; their duration goes in some cases down to the picosecond domain, as it is for some steps in the photochemical cycle of bacteriorhodopsin (Hess *et al.*,

1984). Within such short intervals, the thermal collisions cannot dissipate the stored form of energy and thermodynamics is no longer suited. A detailed older discussion of the difference between thermodynamic and mechanical processes is due to McClare (1971 and 1974) and the return to mechanics as an adequate tool for describing the molecular "machines" was forcefully argued by Blumenfeld (1983).

Finally, it is tempting to observe how the description of processes which account for the main biological roles of the membranes requires different approaches:

i) the long-term function as barries which compartmentalize the living system - equilibrium thermodynamics;

ii) the moderately rapid flows of matter and the energy conversions - nonequilibrium thermodynamics;

iii) the very fast signal reception - non-thermodynamic descriptions.

II.3. Silent and excitable membranes

Coming back to the roles of plasma membrane, they can be formulated as:

i) keeping the inside of the cell a medium adequate for metabolic processes upon ensuring a composition compatible with enzyme functions, providing the substrates for chemical syntheses and disposing of the waste products;

ii) receiving or processing signals from other cells in the body (those belonging to the informational network) or from the environment or, in some cases, generating and transmitting signals to other cells.

The first of these functions is absolutely general, so that the membranes of all living cells must be able to perform it. The second one is also universal but only in the sense that all cells receive signal molecules which command their behaviour as parts of an integrated assembly. These chemical messengers, whose nature vary greatly from one cell to the other, attach to specialized membrane receptors - protein molecules that control specific ionic permeabilities across the membrane. However only some cells are specialized for the rapid transmission of signals, over large distances, either to other cells (this is the case of neurons), or for their own functioning (as it is the case of muscle and gland cells).

While the others are silent, the membranes of these cells are excitable or more precisely electrically excitable, because the speed of propagation cannot be ensured by the mere diffusion of signal molecules, it requiring that the membrane be able to propagate without attenuation electrical signals.

As both silent and excitable membranes have to fulfil the function i), they share the same molecular architecture, with the lipid bilayer preventing the free mixing of hydrophilic molecules and the embedded proteins, some of which ensure the ordered exchanges of matter. When whole macromocules or larger particles have to be imported or exported by endo- and exocytosis, they are envelopped by a certain membrane area which is thus recycled (Pfeffer and Rothman, 1987; Brodsky, 1988). Saying that the lipids are less directly concerned with transmembrane exchanges other than of the lipophilic molecules is not meant to imply that they are passive membrane components, but simply that the timescale of the processes is different from that on which the proteins operate.

From an energetic point of view, of foremost importance are those membrane proteins which directly use the metabolic energy, in the form of ATP or of redox potential differences, to perform the antientropic transport of ions and to give rise to transmembrane gradients of electrochemical potential of some ions, mainly H^+, Na^+, K^+, Cl^- and Ca^{2+}. These proteins are the <u>ion pumps</u> which perform the <u>primary active transport</u> of ions, directly coupled with an exergonic chemical or photochemical reaction.

Primary ion pumps establish concentration gradients which are an intermediate store of metabolic energy. The subsequent entropic flows of the ions along these gradients drive the antientropic transport of other solutes, such as monosaccharides and amino acid molecules, which are thus actively pumped within the cytoplasm by a <u>secondary</u> active transport (also termed coupled transport or cotransport). Out of the many types of ions for which primary ion pumps establish concentration gradients: H^+, Na^+, K^+, Ca^{2+}, Mg^{2+}, Fe^{3+}, Cl^-, HCO^{-3}, only H^+ and Na^+ are found in evolution for driving the secondary active transport of other substances. The H^+-driven active transport occurs in microorganisms and plants, while Na^+ driven flows are specific for animal cells (Kotyk, 1983).

Apart the primary ionic pumps, which establish ionic electrochemical gradients and the carriers which perform the secondary active transport, thus converting one electrochemial gradient into another, a third major class of membrane integral proteins are the <u>conducting sites</u>. These represent pure dissipators, through which the ions flow at high rates, as it was shown in the preceding section, driven by their electrochemical gradients. The common feature of these proteins is that of having two functionally distinct states: an open one, in which the ions are allowed to pass, and a closed one, when they are impermeable.

The dissipators make use of the electrochemical

gradients as the direct source of free energy for the transport, while the pumps rely on the chemical (or the radiant) energy for creating electrochemical gradients. A closer look at the free energy sources of the membrane transport helps establishing a rational classification of the flows and formulating the mechanistic problems to be solved in each case. By now however, the mechanistic aspects are poorly understood and thus subject of debate (Hill, 1983; Tanford, 1983).

The downhill flows, such as the inflow of Na^+ and the outflow of K^+ from the cytoplasm are essentially entropic: they proceed from a higher to a lower concentration until both become equal, or from a higher to a lower electrostatic energy. The free energy change is thus either osmotic: $RT \cdot \ln(c^i/c^e)$, ultimately belonging to the thermal motions, or electrostatic, associated with the weak intermolecular forces: $zF(\Psi^i - \Psi^e)$. When K^+ leaks out of the cytoplasm of a muscle fibre in the extracellular fluid, where its concentration is about 30 times smaller, there is a dissipation of osmotic free energy of ≈ 10.5 kJ/mol. A similar amount of electrostatic free energy is dissipated when Na^+ enters from the outer fluid into the cytoplasm, flowing down an electrical potential difference of about 100 mV, along with the dissipation of about 6 kJ/mol of osmotic free energy, as the concentration of sodium in the cytoplasm is 10 times smaller than outside. In the case of K^+ outflow, the decrease in osmotic free energy is almot matched by the increase in the electrostatic free energy, as the cations move against an electric potential difference.

Obviously, the passive ionic flows are the simplest case of membrane transport. They occur through conducting sites which are transmembrane pathways of low potential energy. These sites are not mere imperfections of the lipid bilayer, but genetically controlled proteic structures (Salkoff and Tanouye, 1986), which play active physiological roles, as for instance ensuring the transmission over large distances of electric signals in nerve fibres. Conceptually the conducting sites may be considered as enzymes because they reduce the electrostatic work required by the transmembrane ionic diffusion from ≈ 250 kJ/mol, to values around 20 kJ/mol. Also they display saturation kinetics at high ion concentration, competitive inhibition and substrate specificity (i.e. the ionic selectivity) (Hille, 1984; Lattorre and Miller, 1983). But the "rates of catalysis" of these unusual enzymes, i.e. the 10^6-10^{19} ions transported per second, are order of magnitude higher than the turnover rates common in biochemistry. Also, the temperature dependence of these rates is very low, the small activation energies showing that there is a fixed hydrophilic region with energy barriers much smaller than those involved when the protein acts as a carrier, whose conformation changes act in concert with the ion transport. The conversion of the osmotic energy into the electrostatic one, as for instance in the case of K^+ ouflow, is easy to conceive because the same miroobjects (the ions) are pushed into a space domain of higher electric potential by their own tendency to escape from where they are agglomerated.

In the case of the active transport of both ions and nonelectrolytes against a concentration gradient, which is a major function of most cell and organelle membranes, the things are considerably less simple.

Practically nothing is known about the molecular properties of the carrier molecules involved in the coupled transport, so that thermodynamics is confined to the limiting statement:

$$\underset{\text{driving ion}}{n \cdot \Delta\tilde{\mu}} \geq \underset{\text{transported solute}}{-\Delta\tilde{\mu}}$$

if n driving ions (H^+ or Na^+) flow downhill for each solute molecule pushed uphill. It expresses the fact that the coupled transport system converts the free energy of the pre-existing activator electrochemical gradient into the free energy of the substrate electrochemical gradient. Assuming that the transported solute (S) is a z-valent ion, the above relation becomes:

$$n[RT \cdot \ln(Na^+_{in}/Na^+_{out})] + F(\Psi^i - \Psi^e) \geq -RT \cdot \ln(S_{in}/S_{out}) - zF(\Psi^i - \Psi^e)$$

for a Na^+ -driven intake of solute [as it is the case of the Na+/aminoacids and Na+/monosaccharides symports in the intestinal epithelial cells (Wright *et al.*, 1983). An important parameter immediately follows, namely the maximum accumulation ratio:

$$(S_{in}/S_{out}) = (Na^+_{in}/Na^+_{out})^n \cdot \exp[-C + z)F(\Psi^i - \Psi^e)/RT]$$

The validity of this equation and of its variants corresponding to different kinetic schemes was repeatedly checked and some deviations discussed (Goffeau and Slayman, 1981; Turner, 1983; Kotyk and Horak, 1985). The unresolved problem is how the transport protein (the carrier) can couple the dissipation in osmotic (and electrostatic) free energy of the driving ion, to transfer an otherwise unaltered solute from low to high concentration. The possibility that the site which binds the solute has a binding constant which depends on the changes in carrier conformation, brought about by the local values of membrane potential (Schwab and Komor, 1978), inevitably remains speculative until getting a precise functional description of the carrier.

A similar, but probably even more complicated problem is in the case of the primary active transport, when the membrane proteins functioning as ionic pump converts the chemical (or the radiant) free energy into electrochemical gradients of one or two ionic species. Ultimately, "chemical" energy also represents electrostatic energy, but belonging to forces at least one order of magnitude stronger than the ionic coulombic forces. The ionic pumps operating in cell membranes are transport ATP-ases which transport uphill the ions at the expense of the free energy of ATP hydrolysis. They are termed "class II ATPases" (Maloney, 1982) and include the (Na^+ + K^+)-ATPase of animal cells (Jürgensen, 1982), the Ca^{2+}-ATPase of the sarcoplasmic reticulum (Schatzmann, 1982), and of plasma membranes, the proton translocating ATPase of fungi (Goffeau and Slayman, 1981) and the gastric H^+/K^+ pump (Sachs *et al.*, 1982). Under physiological conditions they act only as ionic pumps which hydrolyse ATP and go through a cycle of phosphorylation and dephosphorylation steps coupled to conformational transitions.

In the mitochondrial inner membrane and in bacterial membranes, the free energy, derived from redox processes, is converted by the proton translocating FoF1-ATPases into the electrochemical gradients of the protons (Wang, 1983). These ATPases (termed "class I") may work under physiological conditions in either direction of ATP hydrolysis or of its synthesis. A similar class of proton translocating ATPase is in the thylakoid membrane and in green bacterial membranes, utilizing the free energy of the absorbed light quantas. The same input of energy, the relaxation of photon-excited pigment molecules, drives another kind of proton pump in the purple membrane of halobacteria - the bacteriorhodopsin (Stoeckenius *et al.*, 1979).

Most ion pumps are electrogenic as they translocate net electric charge across the membrane. The pump acts, depending on conditions, as a current source or as a voltage source. This transport creates the electrochemical gradients, so that the real primary input of free energy for all membrane transport processes is that one which drives the pumps. During the pumping cycle, the transport proteins undergo conformational changes together with binding and unbinding of ligands, so that it is obvious that the energy supplied by ATP or light is converted in a stepwise fashion into that one of the electrochemical gradients. Since by now the mechanistic properties are incompletely known for all ionic pumps, the microscopic descriptions remain only as models.

Particularly useful appears the idea of representing the pump as a channel with multiple conformational states, i.e. a channel whose energy barrier profile and the binding constant of an ion binding site are transiently modified by the phosphorylation-dephosphorylation cycle or by the light absorption. In a given conformational state, the ion from a binding site has easy access to the cytoplasm, but is prevented by a large energy barrier from escaping to the extracellular medium; in another state, the energy barrier is much higher towards the interior. The idea of such an alternating access is quite intuitive and since a first proposal by Patlak (1957), it is included in most models of active translocation (for extended reviews see Heinz, 1978 (Ch. 6 and 7) and Lèuger, 1984). As concerns the mechanism of free energy coupling in active transport, two more points are worth mentioning.

The turnover cycle of each can be formally written as a reaction catalysed by the pump itself. Thus, in normal conditions, the plasma membrane Na^+/K^+ pump extrudes 3 Na^+ from the cytosol and takes in 2 K^+ for each molecule of ATP hydrolysed (Kyte, 1981):

$$3\,Na^+_{in} + 2\,K^+_{out} + MgATP_{in} \overset{pump}{\rightleftharpoons} MgADP_{in} + P_{in} + 3\,Na^+_{out} + 2\,K^+_{in}$$

The oriented movements of the cations and the exergonic hydrolysis of ATP are coupled obligatorily so that, if the ions flow downhill, the endergonic ATP synthesis takes place. It is unanimously accepted taht the ATP-driven pumps are enzymes, as it is the (Na^++ K^+)ATPase which catalyses the above "reactions". However, as Tanford (1983) pointed out, the transport ATPases are fundamentally different from other enzymes because: 1) they do not function as a rate accelerator of a chemical reaction that could in principle occur (albeit slowly) spontaneously, and 2) the transported molecules do not undergo chemical transformation. Consequently, the close contact between interacting substrates is either unnecessary, or even unplausible. Thus, in these systems, the free energy coupling involves protein-mediated linkage between events occurring some distance apart.

The transduction of free energy may be described on the basis of energy levels of the pump protein in the different states of the pumping cycle (Hill and Eisenberg, 1981). As the turnover rates of the pumps are of the order of $10^2 s^{-1}$ (section II.2), the conformational states of the protein corresponding to the states of the pumping cycle: A<->B<->C...<->A are longlived with respect to the substrates (which differ in the orientation of amino-acid side-chains, in the vibrational modes of the backbone, etc.) between which the transitions occur in the subnanosecond time-range. Accordingly, a pump molecule in a given conformational state (A, B, etc.) may be treated as a chemical quasi-species with a well-defined chemical potential. If $\underline{n}_A$ pump molecules, out of an ensemble of $\underline{n}$, are in the state $\underline{A}$, in view of equation (III.I) the chemical potential of this quasi-species is:

$$\mu_A = \mu^\circ_A + RT\ln (n_A/n)$$

The standard value of the chemical potential, μ°_A, is given by the partition function of the protein in state A, thus being a true molecular quantity. The difference $\mu^\circ_B - \mu^\circ_A$ depends only on the intrinsic molecular properties of the two states and determines how much free energy can be stored in one with respect to the other. On the other hand, the concentration-dependent chemical potential μ_A, is determined by the state of the whole system. The difference $\mu_B - \mu_A$ depends also on the transition rates in the cycle, reflecting both the energy levels of the pump proteins and the kinetic properties of the cycle. The troubling point is that both types of free energy level diagrams, depicting either standard or total values, are beset by the uncertainties concerning the actual intermediary states in the pumping cycles.

Chemo-electrical conversion is quite obvious in the case of excitable cells which generate transient electric currents in response to stimuli. However it occurs - though in less spectacular forms - in all living cells which maintain, at the expense of chemical metabolic energy, steady electric potential differences between the compartments separated by membranes.

The uneven distribution of electric charges in two compartments separated by a membrane gives rise to various kinds of electrical potentials. When the electric field penetrates the whole membrane and can be detected by electrodes introduced in the adjacent bulk solutions, there is a transmembrane potential, while at the boundary between a membrane surface and the corresponding adjacent solution a surface potential can exist. From a thermodynamic point of view, a transmembrane potential is an equilibrium one if the system as a whole attained that state of equilibrium which is possible in the given conditions. Accordingly, an equilibrium potential cannot serve as a source of free energy, unless the conditions are externally changed (for a detailed discussion see Heinz, 1981).

The membranes of all living cells contain ionic pumps, carriers and ionic channels. These are particularly well represented and much studied on the epithelial cells (Van Driessche and Zeiske, 1985), whose property is to maintain distinct ionic composition between the compartments they delineate. In the excitable membranes, the ionic channels have the particular feature of being gated, their structure being so that the transition between the closed and the open conformations is triggered by changes in the transmembrane electric field or, in some specialized zones of contact between communicating excitable cells, by the attachment of specific ligands. Thus, what makes cell excitable is the characteristic of its ionic channels to respond to electrical and chemical stimuli by opening in a transistor-like manner. However such a comparison is not to be further pursued as the opening and closing of the ionic channels are neither solely field-dependent, nor they represent a bi-stable behaviour.

Before a more detailed description of the ionic channels will be given in Chapt. IV, it is to be mentioned the gross distribution of the ionic pumps and channels within the membranes. It was possible to count in situ these membrane proteins because they bind with high affinity specific ligands which can be radiolabelled: the glycoside ouabain binds on the Na/K pump and different neurotoxins on sodium and potassium channels. As indicative orders of magnitude, the Na/K pumps are 10^3 μm^{-2}, while the channels are $(10\text{-}10^2)$ μm^{-2} in unmyelinated axons but 10^3 μm^{-2} in the Ranvier node of myelinated nerve fibres and up to 10^4 μm^{-2} in the neuromuscular junctions (for reviews see Hill, 1984 (Ch. 15) and Vasilescu and Margineanu, 1982, Ch. III).

The sparse distribution of ionic channels and pumps within the unmyelinated axonal membranes explains the rather low proportion of protein in these membranes (Table II.1). Among the protein, they also contain the enzyme acetylcholinesterase (Balerna *et al.*, 1975) and, at least in some cases, acetylcholine receptors similar to those found in the muscle membrane at neuromuscular junctions. In membranes of crustacean axons (unmyelinated), the quantity of protein specifically belonging to sodium channels is 30 x smaller than that of the sodium pump and 7 x smaller than that of acetylcholinesterase.

Table II.1. Composition of squid axon membrane (data quoted by Cotman and Levi, 1975). All numbers are percents of total mass.

	Squid axon membrane
Protein	30
Cholesterol	19
Polar lipids	41
Fatty acids	4
Hydrocarbons	6

Living cells perform all their activities (biosynthetic, mechanical, osmotic, electrical, etc.) at the expense of the free energy derived from the hydrolysis of energy-rich phosphates. No matter how these energy-rich compounds have been obtained (by photophosphorylation, oxidative phosphorylation or by anaerobic glycolysis), the energy to be converted into work is of chemical origin. Another general feature of all living cells is that in their cytoplasm, as well as in the surrounding medium, the enormous amount of water (between at least 50 % and up to 97 % of the mass) makes very many molecules to be ionized. These two conditions are perhaps the ultimate reasons for the ubiquity of chemo-electrical conversion processes, connected with the transport activities of biological membranes.

Both the functions i) and ii) of plasma membranes, as they were formulated at the beginning of this section, have a corresponding bioelectrical expression:

i) the <u>resting potential</u> is the transmembrane electrical potential difference arising as a consequence of the uneven distribution of several ionic species between the cytoplasm and the outside and different permeability characteristics for the various ionic species; it can be detected with intracellular microelectrodes (Ch. III) in all cells, as long as they are "alive", i.e. metabolically active, and shows that the system is out of thermodynamic equilibrium;

ii) the attachment of signal molecules to the membrane receptors induces permeability changes which cause transitory modifications of the resting potential. In excitable cells, specialized for the rapid transmission of electrical signals, an external reduction of the transmembrane potential beyond a given "threshold" value, makes it to continue in a selfmaintained way its decrease and then to change the sign, before returning to the resting value within a few milliseconds. This is the action potential the basic electrical event underlying the nervous and muscular activities. The refinement of the microelectrophysiological techniques and their spread within the last years led to finding that even in non-excitable cells the attachment of ligands on membrane receptors elicit changes of the transmembrane potential but these are considerably slower and they do not propagate without attenuation as it is the case of the action potential. To quote but an example, the extracellular ATP in its tetra-anionic form ATP4- induces depolarization in macrophage plasma membranes, the changes being reversible within minutes or, at least, seconds (Buisman *et al.*, 1988). Thus, in all cases the bioelectric events are a result of the operation of transmembrane channels, the distinction between non-excitable "silent" cells and the excitable ones relying on the specific behaviour of these membrane proteins: when they are able to undergo electric field-mediated transitions between the slow-conducting and the high-conducting states within micro- to milliseconds, the membrane containing them will be excitable. The peculiarity of some cells to respond to a threshold depolarization and propagated reversal of the membrane potential is the direct consequence of the nature and assembly of their ionic channels and understanding the molecular mechanisms of bioelectrogenesis is largely dependent on how close our explanations are to the reality of channel behaviour.

It was already mentioned in §I.3, that the study of bioelectricity is one of the most advanced domains of contemporary biophysics if judged in technical terms, because it benefits from the inherent sensitivity and speed of the electrical measurements, as it will be presented in the next chapter. But it might be a troubling counterpart of this advantage, namely that the prevailing explanations of the bioelectrogenesis are almot exclusively in physical terms, the only connection between bioelectricity and biochemistry being assumed to be the metabolic fueling of the ionic pumps with ATP.

Even before tackling the details, on general grounds, it appears at least intriguing that a major biological process would be only indirectly under the metabolical control. Obviously this monograph is not intended to be neither polemical, nor a plea but its writing was largely prompted by our wish to contribute bridging the non-communicating gap between the biochemical and the biophysical approaches of action potential generation.

Before closing this section, it should be pointed out that now it is firmly established that electrical excitability is not the exclusive property of neurons, muscle cells and certain sensory cells, but it is also shared by a variety of other cell types, such as some protozoa (e.g. Paramecium) and, mainly, by different endocrine cells: the pancreatic islet ßcells, the adenohypophyseal cells, the calcitonin-secreting cells and the adrenal cromaffin cells (Osawa and Sand, 1986). Consequently, the distinction between excitable and non-excitable membranes is somehow imprecise and some membranes considered to belong to the second group appear less silent.

References

Alberts, B. Bray, D., Lewis, J., Raff, M., Roberts, K. and Watson, J.D. (1983) The Molecular Biology of the Cell, Garland Publ., New York (Ch.).
Amos, L.A., Henderson, R. and Unwin, P.N.T. (1982) Prog. Biophys. Mol. Biol. 39, 183-231.
Angelides, K.J., Elmer, L.W., Loftus, D. and Elson, E. (1988) J. Cell Biol. 106, 1911-1925.
Auerbach, A. and Sachs, F. (1984) Ann. Rev. Biophys. Bioeng. 88, 269-274.
Axelrod, D. (1983) J. Membrane Biol. 75, 1-10.
Balerna, M., Fosset, M., Chicheportiche, R., Romey, G. and Lazdunski, M. (1975) Biochemistry 14, 5500-
Benga, G. and Holmes, R.P. (1984) Prog. Biophys. Mol. Biol. 43, 195-257.
Benz, R., Stark, G., Janko, K. and Lafger, P. (1973) J. Membrane Biol. 14, 339-364.
Bishop, W.R. and Bell, R.M. (1988) Ann. Rev. Cell Biol. 4, 579-61O.
Blumenfeld, L.A. (1983) Physics of Bioenergetic Processes (p. 28), Springer, Berlin.
Brodsky, F.M. (1988) Science 242, 1396-1402.
Buisman, H.P. *et al.* (1988) Proc. Natl. Acad. Sci. USA 85, 7988-7992.
Chapman, D. and Hayward, J.A. (1985) Biochem. J. 228, 281-295.
Clément, J.M. (1983) Biochimie 65, 325-338 .
Cook, J.S. (1976) p. 15 in "Biogenesis and Turnover of Membrane Macromolecules" (J.S. Cook, Ed.), Raven Press, New York.
Cornell, B.A. *et al.* (1983) Biochim. Biophys. Acta 732, 473-478.
Cotman, C.W. and Levy, W.B. (1975) p. 185 in "Biochimistry of Cell Walls and Membranes" (C.F. Fox, Ed.) Butterworth, London.
Crowe, J.H. and Crowe, L.M. (1982a) Cryobiology 19, 317-328.
Crowe, L.M. and Crowe, J.H. (1982b) Arch. Biochem. Biophys.

217, 582-587.
Edidin, M. (1981) in Membrane Structure (Finean, J.B. and Michell, R.H., Eds.) p. 37. Elsevier/North Holland, Amsterdam.
Evans, E.A. and Hochmuth, R.M. (1978) Curr. Topics Membr. Transport 10, 1-64.
Evans, E.A. and Skalak, R. (1980) Mechanics and Thermodynamics of Biomembranes p. 86, CRC Press, Boca Raton, Fl. USA.
Finean, J.B., Coleman, R. and Michell, R.H. (1984) Membranes and their Cellular Functions, 3rd edn, p. 163, Blackwell, Oxford.
Finkelstein, A. and Andersen, O.S. (1981) J. Membrane Biol. 59, 155-171.
Franks, F. and Finney, J.L. (eds.) (1981) Hoppe-Seyler's Z. Physiol. Chem. 362, 1177-1198.
Frye, L.D. and Edidin, M. (1970) J. Cell Sci. 7, 319-335.
Goffeau, A. and Slayman, C.W. (1981) Biochim. Biophys. Acta 639, 197-223.
Gorter, E. and Grendel, F. (1925) J. Exp. Med. 41, 439-443.
Graftein, B. and Forman, D.S. (1980) Physiol. Rev. 60, 1167-1283.
Grant, C.W.M. (1984) Can. J. Biochem. Cell Biol. 62, 1151-1157.
Greenblatt, R.E., Blatt, Y. and Montal, M. (1985). FEBS Lett. 193, 125-134.
Guy, H.R. (1984) Biophys. J. 45, 249-261.
Heinz, E. (1978) Mechanics and Energetics of Biological Transport. Springer, Berlin.
Henderson, R. and Unwin, P.N.T. (1975) Nature (London) 257, 28-32.
Hess, B., Kuschmitz, D. and Engelhard, M. (1984) in Information and Energy Transduction in Biological Membranes, p. 81, Alan R. Liss, New York.
Hill, T.L. (1983) Proc. Natl. Acad. Sci. USA 77, 1681-1683.
Hill, T.L. and Eisenberg, E. (1981) Q. Rev. Biophys. 14, 463-511.
Hille, B. (1984) Ionic channels of excitable membranes (Sinauer Associates Inc., Sunderland Mass.)
Hubbell, W.L. and McConnell, H.M. (1968) Proc. Natl. Acad. Sci. USA, 61, 12-16.
Hubbell, W.L. and McConnell, H.M. (1969) Proc. Natl. Acad. Sci. USA 64, 20-27.
Jackson, M.B. and Sturtevant, J.M. (1978) Biochemistry 17, 911-915.
Jakobson, K., Ishihara, A. and Inman, R. (1987) Ann. Rev. Physiol. 49, 163-175.
Jorgensen, P.L. (1982) Biochim. Biophys. Acta 694, 27-68.
Karlish, S.J.D. and Pick, N. (1981) J. Physiol. (London) 312, 505-529.
Kotyk, A. (1983) J. Bioenerg. Biomembr. 15, 307-319.

Kotyk, A. (1985) in Water and Ions in Biological Systems (Pullman, A., Vasilescu, V. and Packer, L., eds.), p. 463. Union of Societies for Medical Sciences, Bucharest.
Kotyk, A. and Janacek, K. (1977) Membrane Transport, Ch. 2, pp. 169-181, Plenum, New York.
Kotyk, A. and Horak, J. (1985) in Water and Ions in Biological System (Pullman, A. Vasilescu, V. and Packer, L., eds.), p. 343, Plenum Press, New York.
Latorre, R. and Miller, C. (1983) J. Membrane Biol. 71, 11-30.
Lafger, P. (1984) Biochim. Biophys. Acta 779, 307-341.
Lenaz, G. (1984) in Biomembranes: Dynamics and Biology (Guerra, F.C. and Burton, W.M., eds.) p. 111, Plenum Press, New York.
Lennarz, W.J. (ed.) (1980) The Biochemistry of Glycoproteins and Proteoglycans, p. 241, Plenum Press, New York.
McClare, C.W.F. (1971) J. Theor. Biol. 30, 1-34.
McClare, C.W.F. (1974) Ann. N.Y. Acad. Sci. 227, 74-97.
McCloskey, M. and Poo, M.-m. (1984) Int. Rev. Cytol. 87, 19-81.
Maloney, P.C. (1982) J. Membrane Biol. 67, 1-12.
Marchesi, V.T. (1985) Ann. Rev. Cell Biol. 1, 531-561.
Margineanu, D.G. (1987) Arch. Int. Physiol. Biochim.95, 381-422.
Neher, E. and Stevens, C.F. (1977) Ann. Rev. Biophys. Bioeng. 6, 345-381.
Op den Kamp, J.A.F. (1979) Ann. Rev. Biochem. 48, 47-71.
Ozawa, S. and Sand, O. (1986) Physiol. Rev. 66, 887-952.
Patlak, C.S. (1957) Bull. Math. Biophysics 19, 209-235.
Pfeffer, S.R. and Rothman, J.E. (1987) Ann. Rev. Physiol. 56, 829-
Pullman, A., Vasilescu, V. and Packer, L. (eds.) (1985) Water and Ions ins Biological Systems. Plenum, New York.
Pumplin, D.W. and Fambrough, D.M. (1982) Ann. Rev. Physiol. 44, 319-335.
Quinn, P.J. (1981) Prog. Bioph. Molec. Biol. 38, 1-104.
Ritchie, J.M. and Rogart, R.B. (1977) Proc. Natl. Acad. Sci. USA 74, 211-215.
Sachs, G., Wallmark, B., Saccomani, G., Rabon, E., Stewart, H.B., DiBona, D.R. and Berglindh, T. (1982) Curr. Top. Membr. Transp. 16, 135-160.
Salkoff, L.B. and Tonouyje, M.A. (1986) Physiol. Rev. 66, 301-
Sandermann, H. (1983) Membranbiochemie, p. 9, Springer, Berlin.
Schatzmann, H.J. (1982) in Membrane Transport of Calcium (Carafoli, E., ed.) p. 41, Academic, New York.
Schein, S.J. (1976) J. Exp. Biol. 65, 725-736.

Schimke, R.T. (1975) p. 229 in Biochemistry of Cell Walls and Membranes (C.F. Fox, ed.) Butterworth, London.
Schoenheimer, R. *et al.*, (1942) The Dynamic state of body constituents. Harvard University Press, Cambridge, Mass.
Schramm, M. and Selinger, Z. (1984) Science (Washington) 225, 1350-1356.
Schwab, W.G.W. and Komor, E. (1978) FEBS Lett. 87, 157-160.
Siekevitz, P. (1975) in Cell Membranes: Biochemistry Cell Biology and Pathology (Weissmann, G. and Clairborne, R., eds.) p. 115, HP Publ. Co., New York.
Shinitzky, M. (1984) Biochim. Biophys. Acta 738, 251-261.
Shintizky, M. and Henkart, P. (1979) Int. Rev. Cytol. 60, 121-147.
Singer, S.J. (1974) Ann. Rev. Biochem. 43, 805-833.
Stoeckenius, W., Lozier, R.H. and Bogomolni, R.A. (1979) Biochim. Biophys. Acta Z505, 215-278.
Tanford, C. (1983) Ann. Rev. Biochem. 52, 379-409.
Thompson, T.E. and Huang, C. (1985) in Physiology of Membrane Disorders (Andreoli, T., Fanestil, D.D., Hoffman, J.F. and Schultz, S.G., eds.) 2nd ed., p. 25, Plenum Press, New York.
Turner, R.J. (1983) J. Membrane Biol. 76, 1-15.
Ullrich, A. *et al.* (1985) Nature 313, 756-761.
Van Driessche, V. and Zeiske, W. (1985) Physiol. Rev. 65, 833-903.
Vasilescu, V. and Margineanu, D.G. (1982) Introduction to Neurobiophysics, Abacus, Turnbridge Wells.
Wang, J.H. (1983) Ann. Rev. Biophys. Bioeng. 12, 21-34.
Welte, W. and Kreutz, W. (1982) Biochim. Biophys. Acta 692, 479-488.

CHAPTER III
PHENOMENOLOGICAL ASPECTS OF BIOELECTRICITY

Chemo-electrical conversion is quite obvious in the case of excitable cells, which generate transient electric currents in response to stimuli. However it occurs - though in less spectacular forms - in all living cells, which maintain, at the expense of chemical metabolic energy, steady electric potential differences between the compartments separated by membranes.

Among the fundamental characteristics of all living cells is the existence of an electric potential drop between the external and the internal sides of the plasma membrane. As both cytoplasm and interstitial fluid are media with high electrical conductivity , the electric potential is the same at any point within them, the potential drop occuring only across the membrane. This transmembrane potential difference, present in all metabolically active cells is the resting potential (RP). Initially, it was called "injury potential" owing to the way by which it was detected, upon cutting a bundle of muscle fibres and applying one electrode on the injured surface and the other on the intact one. Always the injured area, i.e. the inside of the cells, appears negatively charged with respect to the intact one, but because of the inevitable short-circuiting between the two zones of the tissue, the injury potential is only a fraction of the true RP.

III.1. Resting potential of the cells

A landmark in the progress of cell physiology in general and of biophysics of excitable cells in particular was represented by the development (by G.N. Ling and R.W. Gerard in 1949) of intracellular glass microelectrodes. These are glass micropipettes with the tip of less than 0.5 μm (Fig. III.1a). When piercing the cellular membrane, the micropipette does not induce any significant short-circuit between cytoplasm and the extracellular fluid, because the interfacial tension makes the membrane to shrink around its tip.

The micropipette is filled with a 3 M KCl solution, near saturated, in order to reduce as much as possible the internal resistance of the electrode which is large because of the very small cross-section. Even if this solution with high conductance is used, the resistance of glass miroelectrodes amounts to 10 MΩ, which means that the amplifiers used for measuring the biopotentials must also have a very large input resistance. KCl solution is universally used since both K^+ and Cl^- are non-toxic (as they normally are present within the cytoplasm) and both have the same mobility, so that no junction potential appears at their contact with the cytoplasm.

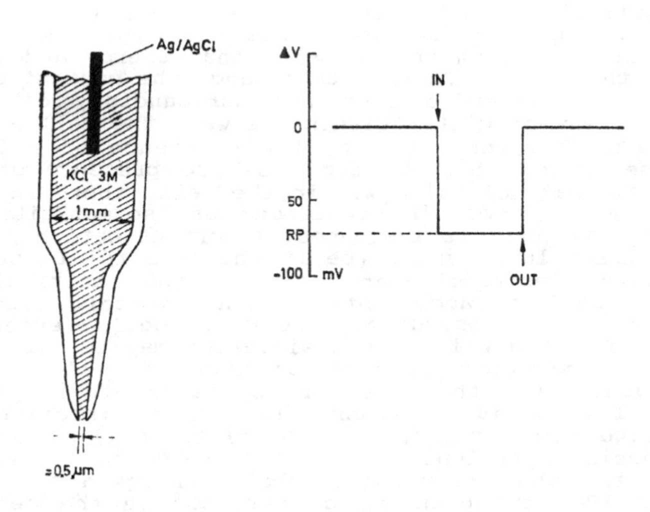

Figure III.1. Schematic drawings of:
a) glass micropipette used as an intracellular electrode;
b) potential difference between a reference electrode immersed in the external solution and a microelectrode at the entrance and the exit from a cell (RP is the resting potential).

Between the intracellular microelectrode and a reference one (for example a calomel electrode) left in the external solution, a potential difference appears, the cell inside always being negative with respect to the outside, generally between - 50 - and - 100 mV - (Fig. III.1b).

There is an obvious connection between the resting potential (RP) and cell metabolism because all the factors depressing the metabolism, such as the lack of oxygen, the presence of general metabolic inhibitors and the decrease in temperature cause a reduction in the RP value till its total disappearance. This is because of the metabolic processes, which maintain an ionic composition quite different in the cytoplasm and the extracellular fluid, the resting potential being the electrical expression of this asymmetry. As it was previously mentioned (§II.1), one of the first characteristics that must have appeared early in the chemical evolution leading to biological systems is the isolation of a metabolic volume segregated from the rest of the environment. This implies that the reaction volume be limited by a membrane having specific properties of semi-permeability. By its very nature, such a system, even rather primitive, is faced with a problem of volume regulation. By definition the system must be thermodynamically open in the sense that energy and matter must flow through it. On the other hand, the content of the cell must be different from its surrounding, otherwise there is no way of distinguishing between the cell and the environment. Therefore, due to the existence of a DONNAN effect (see below) brought about by the presence of non-permeant charged molecules within the cell, there is a net inflow of water, even in conditions of isosmoticity. Na^+ ions being the quantitatively most important cation present in the extracellular space (be it the primordial ocean or contemporary extracellular fluid in multicellular organisms) tend to accumulate within the cell, together with water. As a consequence, the cell swells, eventually up to the bursting point. This situation may be avoided if Na^+ ions are continuously expelled from the cell interior. It is indeed found that most of the cells studied so far keep a low sodium content in their interior, by transporting this ionic species out of the cell against its electrochemial gradient. Obviously this is a process requiring the expenditure of metabolic energy and today we know that ATP, the main energy currency in the cell, is used up in the process. The transduction system is the enzyme (Na^+ + K^+)-ATPase, a tetrameric protein that spans across the membrane. The hydrolysis of ATP induces a change in the configuration of the enzyme and as a result Na^+ ions are transported out of the cell in exchange for K^+ ions. Consequently part of the free energy of ATP hydrolysis is stored as ionic gradients since the cell interior is highly concentrated in K^+ ions and has a low-sodium content, as compared to the extracellular fluid (Table III.1).

III.1.a. Equilibrium transmembrane potentials

The uneven distribution of electric charges in two compartments separated by a membrane gives rise to various kinds of electrical potentials. When the electric field penetrates the whole membrane and can be detected by electrodes introduced in the adjacent bulk solutions, there is a transmembrane potential, while at the boundary between a membrane surface and the corresponding adjacent solution a surface potential can exist. From a thermodynamic point of view, a transmembrane potential is an equilibrium one, if the system as a whole attained the state of equilibrium possible in the given conditions. Accordingly, an equilibrium potential cannot serve as a source of free energy, unless the conditions are externally changed (for a detailed discussion see Heinz, 1981).

Table III.1. Concentrations values for the most representative ions within the cytoplasm of three types of cells and in their extracellular fluids, together with the corresponding electrochemical equilibrium potentials of these ions (ENa, EK and ECl) and the resting potential of the cells.

Cell	Type of ion	Ionic concentration. (mM)		E_{Na}; E_k; E_{Cl} (mV)	Resting poten-tials (mV)
		Intra cellular	Extra cellular		
Squid giant axon	Na^+ K^+ Cl^-	45 410 40	440 22 560	+ 57 - 74 - 66	- 61
Frog muscle fibre	Na^+ K^+ Cl^-	13 138 3	110 2.5 90	+ 55 - 110 - 86	- 99
Human red blood cell	Na^+ K^+ Cl^-	19 136 78	155 5 112	+ 53 - 83 - 9	- 10

Let there be two solutions of the same salt but of different concentrations, separated by a cation selective membrane, for instance one with pores sufficiently narrow and densely lined with fixed negative charges so that to prevent the passage of anions.

The system will reach a state of equilibrium in which not only the impermeant anion, but also the permeant cation, indicated by the script (+) will cease to flow. This situation (Fig. III. 2a) is defined by the absence of any difference in electrochemical potential for the permeant ion in the two compartments.

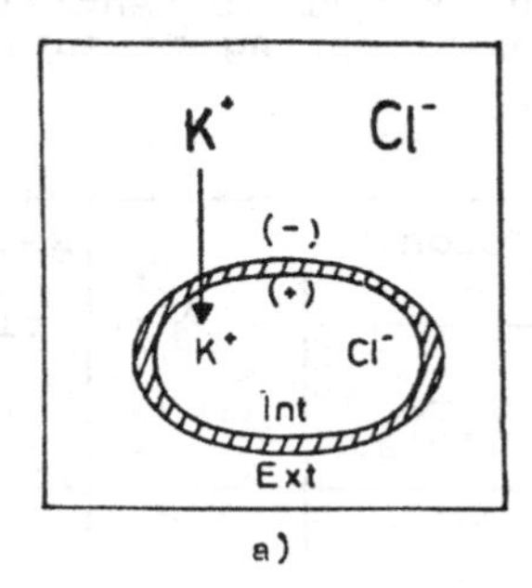

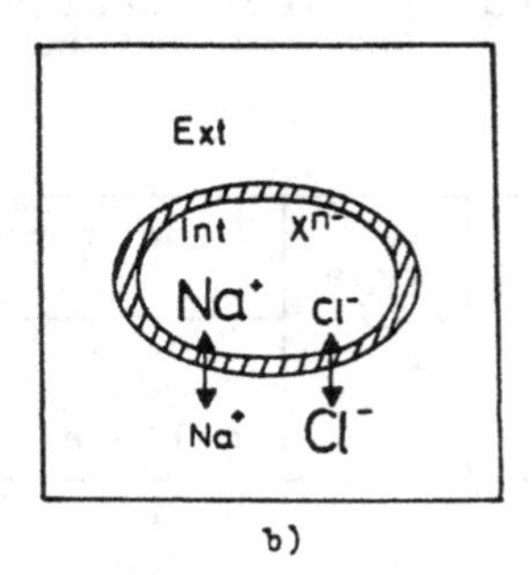

Figure III.2. Equilibrium distributions of ions.
a) When an ideal cation selective membrane separates a more diluted internal salt solution from a more concentrated external one, at equilibrium the interior is positively charged,b) when the membrane is permeable to microions irrespective of their charge, but is nonpermeable to a poly-anion (X^{n-}) confined to the interior, at equilibrium the interior is negatively charged..

The transmembrane electrical potential difference, called in this case Nernst electrochemical equilibrium potential is:

$$E_+ \equiv \Psi^i - \Psi^e = \frac{RT}{Z+F} \ln \frac{C_+^e}{C_+^i} \qquad \text{(III.1)}$$

This equation can be rewritten in the form:

$$C_+^e = C_+^i \exp\left(-z_+ F e_+ / RT\right) \qquad \text{(III.2)}$$

showing that the ions are distributed according to Boltzmann's law for molecular assemblies at equilibrium.

In the process of equilibration, the permeant cation enters the more diluted compartment, until this becomes positively charged, so that the repulsive electrical field exactly compensates the concentration gradient and the further diffusional transfer is prevented.

Nernst equilibrium potential is important for theoretical reasons, as it is the simplest case of an equilibrium distribution of ions and because it allows to express a concentration ratio by means of an equivalent electrical potential. However, such a distribution is not encountered in living cells, because the biological membranes are not ideally selective but they allow the passage - though with different permeabilities - of many small ions.

A more realitic case of equilibrium distribution describes the situation depicted in Fig. III.2.b: a large non-permeant anion is restricted to the internal compartment, while a small cation and a small anion move freely between the two solutions. In this case the equilibrium distribution of ions is established by two independent conditions: a) no electrochemical potential gradient should exist for any of the permeant ions:

$$\bar{\mu}^{i}_{Na} = \bar{\mu}^{e}_{Na} \qquad \text{and} \qquad \bar{\mu}^{i}_{Cl} = \bar{\mu}^{e}_{Cl}$$

b) both compartments must be electroneutral, which implies:

$$C^{i}_{Na} = C^{i}_{Cl} + nC_{x} \qquad \text{and} \qquad C^{e}_{Na} = C^{e}_{Cl}$$

where n is the charge of the polyanion.

The equilibrium condition a) gives the electrical potential difference (The Donnan potential):

$$\Psi^{i} - \Psi^{e} = \frac{RT}{F} \ln \frac{C^{e}_{Na}}{C^{i}_{Na}} = \frac{RT}{F} \mathrm{Ln} \frac{C^{i}_{Cl}}{C^{e}_{Cl}}$$

The two equal concentration ratios:

$$r = (C^{i}_{Na} / C^{e}_{Na}) = (C^{e}_{Cl}/C^{i}_{Cl})$$

are called "Donnan ratio". The electroneutrality condition b) leads, after simple manipulations, to obtain this ratio:

$$r = [\, nC_{x} + \sqrt{4(C^{e}_{Cl})^{2} + (nC_{x})^{2}}\,]/2\, C^{e}_{Cl}$$

and the Donnan potential can be written:

$$\Psi^{i} - \Psi^{e} = -\frac{RT}{F} \ln r$$

The presence in the inner compartment of the nonpermeant polyanion implies the net transfer into this compartment of an amount of permeant cation and thus of osmolarity. Accordingly, an associated osmotic water inflow occurs until the increment in osmotic pressure keeps the water at equilibrium.

The Donnan potential is a true equilibrium one and it is fully relevant for describing the ionic distributions in systems separated by passive membranes, whose permeability properties are solely defined by their porosity. However, for living cell membranes, a far more complex situation holds.

At the turn of this century, J. Bernstein stated that RP represents the electrochemical equilibrium potential of potassium, since the cell membrane would be selectively permeable to this ion only. This means that the electric work necessary to transport one mole of potassium against the potential difference Ek that is F. Ek (where Faraday's number F = 96,500 C is the electric charge of one mole of monovalent ions) must be equal to the osmotic work done when that mole passes from the higher internal concentration to the smaller external one:

$$RT \cdot \ln(C_k^e) - RT \cdot \ln(C_k^i)$$

(where R is the gas constant and T the absolute temperature). From this relationship:

$$F \cdot E_k = RT \cdot \ln(C_k^e) - RT \cdot \ln(C_k^i) \quad RT \cdot \ln(C_k^i)$$

we get:

$$E_k = \frac{RT}{F} \cdot \ln \frac{C_k^e}{C_k^i} \qquad \text{(III.1')}$$

This formula was established by W. Nernst as the general case for any electric cell in which a difference of electrical potential occurs, owing to a difference in the concentration of an ion, and was used by Bernstein for explaining the RP.

The electrochemical equilibrium potentials for K^+, Na^+ and Cl^- calculated with Nernst's formula, are given in Table III.1. Upon comparing these values with the measured RP, it was found that, while the ionic distribution of sodium corresponds to an equilibrium potential completely different from the RP (e.g. + 55 mV against - 61 mV), for both potassium and chloride, the distribution approximately corresponds to the equilibrium condition. This kind of study showed that the variations of Cl^- concentration in the external solution as well as the variations of Na^+ concentration have a rather small influence on the resting potential. If, on the other hand, K^+ concentration is changed, an obvious logarithmic variation of the type predicted by Nernst's formula will be found.

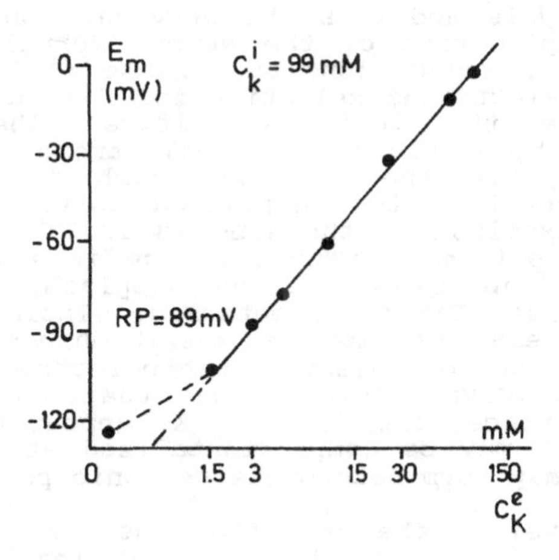

Figure III.3 Dependence of the membrane potential E of an amphibian neuroglia on the external potassium concentration (C^e_K). Notice the log scale on the abscissa and the deviation from linearity for small C^e_K values (from Kuffler et al., 1966).

Of the numerous investigations confirming this fact, Fig. III.3 shows the results of Kuffler *et al.*, (1966) for the variation of membrane potential in the leech neuroglia as a function of K^+ concentration in the external solution.

In Nernst's formula, the electrochemical equilibrium potential is calculated on the basis of ion concentrations, as measured by chemical or spectroscopic methods. However, strictly speaking, the activities of the ions should be considered, i.e. the products of concentration and an activity coefficient expressing the interaction forces of that ion in solution: $a_i=c_i f_i$. A high degree of binding of the respective ion, within structures preventing its free diffusion, is reflected by a low activity coefficient.

Within the "association-induction" theory, Ling (1962; 1977) maintains that hydrated potassium ions are preferentially absorbed, rather than sodium ions, in the meshes of a cytoplasmic lattice of three-dimensionally structured protein molecules, a lattice into which they can be accomodated owing to their smaller radius. Such a preferential binding of potassium would imply that its activity is much smaller than the concentration established by chemical analysis and thus the previous considerations based on the application of the Nernst formula would no longer be valid. However, long ago, Hinke (1961), working with glass microelectrodes selective for Na^+ and K^+, could directly measure the ionic activities. The electric potential "seen" by a selective microelectrode depends just on the activity of the ion to which the glass is selectively permeable. Hinke confirmed that the activity coefficient in axoplasm is the same as in Ringer solution and, even more, pointed out that sodium ions are somehow "bound" or masked up to 25 % in the axoplasm, contrary to the assumption in Ling's hypothesis. Incidentally we mention that several studies performed on various cells using different types of selective microelectrodes (Lev and Armstrong, 1975; Civan, 1978) show that only a small percentage of intracellular Na^+ and K^+ appear bound, even though these ions may be compartmentalized at subcellular level, and thus may form heterogeneous ionic populations in the cytoplasm.

In conclusion, the use within Nernst's formula of the ionic concentrations is justified at least as a fair approximation. But, when considering the magnitude of the resting potential, both the data in Table III.1 and that in Fig. III.3 reveal deviations from the predicted values.

The consideration of RP as the equilibrium potential of K^+ is based on the hypothesis that the cell membrane would be impermeable to Na^+, as well as to other ions. However, since the early fifties, radiotracer studies have shown that plasma membranes are permeable to various ions, particularly to Na^+ and Cl^-, though with a lesser permeability than for K^+.

III.1.b. Steady-state nonequilibrium transmembrane potential

The tracer studies imposed the conclusion of paramount theoretical importance that the resting state of the cells does not correspond to any equilibrium distribution of ions but to a nonequilibrium steady state in which the passive ionic flows, downhill their electrochemical gradients, are balanced by opposite active flows driven by the membrane ionic pumps. Accordingly, the RP has to be defined as that transmembrane potential difference at which there is no net electric current. This steady-state condition is self evident for if a net electric current would exist, a permanent accumulation of electric charge in one of the two compartments, separated by the membrane, would occur, which is physically and biologically senseless.

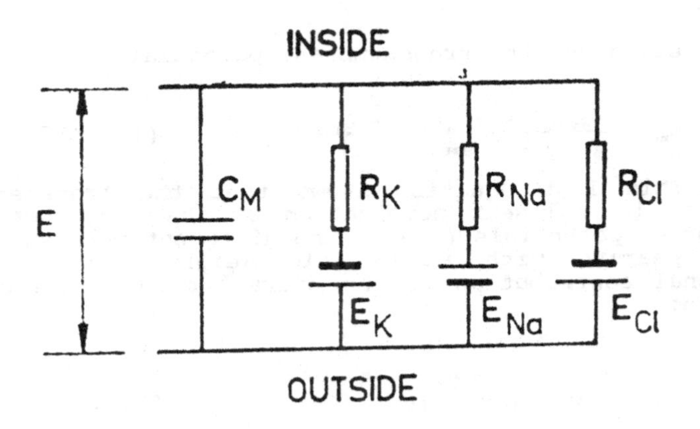

Figure III.4. The electric circuit equivalent to a given area of a plasma membrane: R_K, R_{Na} and R_{Cl} are the partial membrane resistances for each type of ion; K_K, E_{Na} and E_{Cl} are their corresponding Nernst potentials; C_m is the membrane capacity and E_m is the transmembrane potential drop.

The electrical properties of plasma membranes can be represented as in Fig. III.4 by three kinds of components: the dissipative elements (the resistances), the active elements (the batteries representing the uneven concentrations inside and outside) and the reactive element C_m, associated with the dielectric properties of the membrane lipid bilayer. According to Ohm's law, the current through each resistance is proportional with the driving force $E_m - E_i$ (where i = K, Na and Cl) and vanishes when $E_m = E_i$. Thus:

$$I_K = (E_m - E_K)/R_K \quad \text{or} \quad I_K = g_K (E_m - E_K) \qquad (III.2)$$

where $g_K = 1/R_K$ is the partial conductance of the membrane for K^+ ions. Similar expressions hold for Na^+ and Cl^-:

$$I_{Na} = g_{Na}(E_m - E_{Na}) \text{ and } I_{Cl} = g_{Cl}(E_m - E_{Cl}).$$

The steady-state condition:

$$I_K + I_{Na} + I_{Cl} = 0 \qquad (III.3)$$

immediately gives the transmembrane potential:

$$E_m = \frac{g_K E_K + g_{Na} E_{Na} + g_{Cl} E_{Cl}}{g_K + g_{Na} + g_{Cl}} \qquad (III.4)$$

This last equation shows that the transmembrane potential is a linear combination of the electrochemical equilibrium potentials (E_i) of the different permeant ions, each appearing with a specific weight equal to its fractional contribution to the overall conductance of the membrane:

$$E_m = \sum_{i=1}^{n} g_i E_i \Big/ \sum_{i=1}^{n} g_i \qquad (III.4')$$

The equation (III.4) and its more general form III.4' are well suited for accounting the measured values of the RP and remarkably simple to interpret in macroscopic electrical terms. However, they almost canceal the origin of the transmembrane potential, which is the electrodiffusion across the membrane of the different permeant ions, driven by the spatial gradients of their electrochemical potentials. In the general parlance of non-equilibrium thermodynamics, these represent the forces X_i which produce the ionic flows J_i.

If, for simplifying the presentation without loosing any physical generality, one considers the variations along only one direction:

$$X_i = d\tilde{\mu}_i/dx = -\left(\frac{RT}{C_i}\frac{dC_i}{dx} + z_iF\frac{d\Psi}{dx} \right)$$

(all the notations were previously introduced, the subsript i denotes as before the different ionic types, and the minus sign indicates the entropic tendency of the decreasing of gradients).

By definition, the flow Ji is the amount of mass (m_i) carried by those ions in unit time throuh the unit of surface (S):

$$J_i = (\Delta m_i/\Delta t)/S = C_i v_i$$

where v_i is the migration velocity of the ions, in turn proportional with the driving force: $v_i = u_i X_i$. The proportionality factor u_i is the mobiliy of the ions within the membrane. Accordingly:

$$J_i = -RTu_i\left[\frac{dc_i}{dx} + \frac{Z_iFC_i}{RT}\left(\frac{d\Psi}{dx}\right)\right] \qquad (III.5)$$

this being the renewed Nernst-Planck electrodiffusion equation.

In 1943, D.E. Goldman, then in 1949 A.L. Hodgkin and B. Katz, applied a method (originally used by Nevill Mott for the conduction of electrons at metal semiconductor junctions) to integrate the equation (III.5) across the membrane of thickness l. A simple manipulation leads to:

$$J_i = -[RTu_i/\exp(z_if\Psi/RT)] \times \frac{d}{dx}.[C_i\exp(z_iF\Psi/RT)]$$

If the partition coefficient of the ions i between the membrane and the adjacent solutions (where the concentrations are $C^e{}_i$ and $C^i{}_i$) is β_i, it means that just inside the edges of the membrane, the concentration are $\beta_i C^i{}_i$ at x = 0 and $\beta_i C^e{}_i$, at x = l. Obviously, the condition of mass conservation ensures that J_i does not depend on x, so that the integration of the above equation from x = 0 to x = l gives:

$$J_i = \int_{x=0}^{l} \exp(z_iF\Psi/RT)dx = -\beta_iRTu_i[C_i^e \exp(z_iF\Psi^e/RT - C_i^i\exp(ziF\Psi^i/RT]$$

If now one makes the simplifying assumption that <u>the electric field across the membrane is constant</u>, i.e.:

$$- \frac{d\Psi}{dx} \equiv \frac{E_m}{l} \qquad \text{(III.6)}$$

it follows:

$$\Psi(x) = \Psi^i - \frac{E_m}{l} x$$

and the integral in the left side is straightforward, so that:

$$J_i = \frac{z_i F \beta_i U_i}{[\exp(-z_i F E_m/RT)-1]} \left(\frac{E_m}{l}\right) [C_i^e \exp(-z_i F E_m/RT) - C_i^i]$$

The electric current carried by a molar flow J_i of ions with valence zi is: $I_i = z_i F J_i$ and the identification of the first term in the right side of equation III.5 with a simple diffusion equation, allows introducing the <u>permeabilities</u> P_i of the membrane for each type of ions.

The presumed linear proportionality between the diffusional flow J^{dif}_i and the concentration difference implies:

$J_i^{dif} \approx P_i(C_i^i - C_i^i)$ and the diffusion term in the Nernst-Planck equation III.5 is

$$J_i^{dif} = \frac{RTU_i\beta_i}{l}(C_i^e - C_i^i)$$

again with the assumption of linearity:

$$- \frac{dC_i}{dx} = \frac{\beta_i C_i^e - \beta_i C_i^i}{l} \qquad \text{(III.7)}$$

Upon comparing the two expressions of the diffusional flow, the permeabilities appear as combinations of mobilities, partition coefficients and the thickness of the membrane:

$$P_i = \frac{RTU_i\beta_i}{l}$$

With these, the electric currents respectively carried by K+, Na+ and Cl- ions are:

$$I_K = \frac{F^2E_m}{RT} \cdot \frac{P_K\,[C_K^e \exp(-FE_m/RT) - C_K^i]}{\exp(-FE_m/RT) - 1}$$

$$I_{Na} = \frac{F^2E_m}{RT} \cdot \frac{P_{Na}\,[C_{Na}^e \exp(-FE_m/RT) - C_{Na}^i]}{\exp(-FE_m/RT) - 1}$$

$$I_{Cl} = \frac{F^2E_m}{RT} \cdot \frac{P_{Cl}\,[C_{Cl}^e \exp(-FE_m/RT) - C_{Cl}^i]}{\exp(-FE_m/RT) - 1}$$

$$= \frac{F^2E_m}{RT} \cdot \frac{P_{Cl}\,[C_{Cl}^i \exp(-FE_m/RT) - C_{Cl}^{ei}]}{\exp(-FE_m/RT) - 1}$$

and the steady-state condition (III.3) gives:

$$(P_K C_K^e + P_{Na} C_{Na}^e + P_{Cl} C_{Cl}^i) \cdot \exp(-FE_m/RT) = P_K C_K^i + P_{Na} C_{Na}^i + P_{Cl} C_{Cl}^e$$

or:

$$E_m = \frac{RT}{F} \ln \frac{P_K C_K^e + P_{Na} C_{Na}^e + P_{Cl} C_{Cl}^i}{P_K C_K^i + P_{Na} C_{Na}^i + P_{Cl} C_{Cl}^e} \qquad \text{(III.8)}$$

The equation III.8, which in fact is identical with III.4, is the standard Goldman-Hodgkin-Katz (GHK) expression for the transmembrane resting potential of the cells. The agreement of the experimentally measured values of RP with those predicted by equation III.8 for various concentrations in the external solution is much better than that with the values given by equation III.1′ and it was actually established in the case of different cells.

It may easily be seen that GHK formula will reduce to Nernst's simpler one, if both P_{Cl} and P_{Na} are much smaller than P_K, which explains the previously discussed fact that RP relates logarithmically to the external concentration of potassium and is less influenced by variations of the external concentration of sodium. Although the derivation of GHK equation (III.8) embodies the simplifying linearity assumptions (III.6) and (III.7), it is conceptually much closer to the dynamic reality of the living cells, because it deals with nonequilibrium steady-states, instead of a purely hypothetical equilibrium. Both (III.4) and (III.8) equations show that the membrane potential Em tends to approach the electrochemical equilibrium potential Ei of that ion for which the membrane conductance is much higher than for the others.

The GHK formula is used for determining the ratios of ionic permeabilities when ionic concentrations in the cell inside and outside are known: in many cases, such as those of molluscan neurons or frog muscle, either the contribution of chloride to RP is negligible, or it is possible to work in solutions without chloride, so that only the contributions of potassium and sodium remain. Also, because $P_{Na} < P_K$ and $C^i_{Na} << C^i_K$, the term $(P_{Na}/P_K)C^i_{Na}$ may be equally ignored, so that:

$$RP \approx \frac{RT}{F} \ln \frac{C^e_K + (P_{Na}/P_K)C^e_{Na}}{C^i_K}$$

On the basis of this relation, the permeability ratio P_{Na}/P_K can be easily inferred or, if this ratio is known from flux measurements, the internal concentration of potassium can be estimated.

The representation of the electrical characteristics of cell membranes as separate batteries and resistances for each ion (as they appear in Fig. III.4) is based on the idea that in the membrane separate ionic channels exist - a fact confirmed by the possibility of selectively blocking the electric currents carried by sodium and potassium through the axon membrane - and on the independence principle of ionic flows (Hodgkin and Huxley, 1952), according to which each ionic flow is influenced solely by the electrochemical gradient of that ion. On this kind of representations are based the descriptions of the initiation and propagation of excitation in axonal membranes, as well as in other types of excitable membranes, including the synaptic ones and the membranes of receptor cells.

III.2. The passive propagation of potential changes. Axons as electric cables

As it already appeared from Fig. III.4, in electrical terms, every cell membrane has resistive properties, which express its ionic permeability and a capacitive behaviour, arising from the dielectric properties of lipid bilayer. When a square pulse of current is fed into a cell via a microelectrode, the applied current first charges (or discharges) the membrane capacitance, then flow as ionic current. The membrane potential is changed during the whole process. That component of the current affecting first the capacitance of the membrane is dependent on the excess charges that are applied. An influx of positive charges reduce the negative charge of the inner face of the cell membrane. Therefore the positive charge on the outer side of the membrane decreases also and the amount of positive charges released in the extracellular solution is obviously equal to the number of charges fed in the system to decrease the negativity of the inner side of the membrane. Thus there is a current flow through the membrane, though actually no carriers of charge have crossed the membrane. This current, generated by charge displacement via the membrane capacitance, is the capacitive current, I_c.

III. 2.a. Membrane time and length constants

Given a constant supply of charge, the charge of the membrane capacity together with the membrane potential should change at a constant rate. This is not what one observes (Fig. III.5.a) since the rate of change decreases with time.

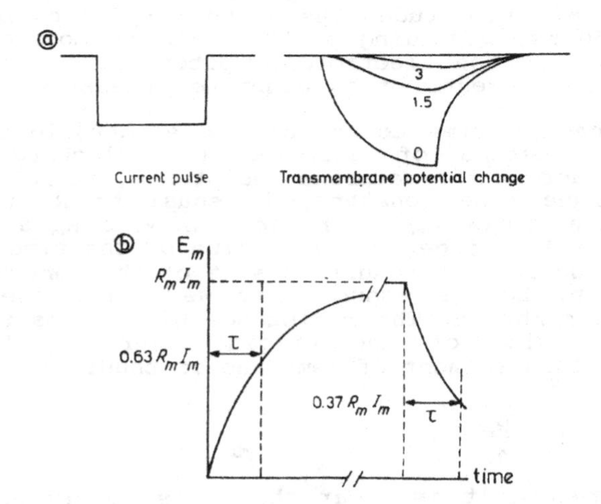

Figure III.5. Evolution of transmembrane potential E_m in response to a square pulse of injected current. In a) the potential changes recorded at 0; 1.5 and 3 times the length constant away from the current passing electrode are represented and in b) the time course of E_m is detailed, to define the time constant τ .

This is due to the fact that, in resting conditions, the membrane is permeable to K^+ and Cl^- ions and to a lesser extent to Na^+ ions. Ionic currents are flowing through the membrane and, provided that the resting potential is held constant, the algebraic sum of these ionic currents is zero. By applying a constant supply of electric charges to the membrane, its electric potential varies and a net ionic current is generated that is proportional to the magnitude of the shift in potential. As depolarization proceeds, less and less current is available for discharging the membrane capacitance and the membrane potential changes more and more slowly as time proceeds. Finally it remains constant and the whole current is flowing through the membrane resistance as ionic current, I^i (Fig. III.5.b).

Therefore, though we apply a square pulse through the membrane, for the reasons stated above we record a signal, the so-called electrotonic potential, which shows exponential rise and decline characterized by a time constant τ , i.e. the time required for the potential to rise to (1-1/e) of its final amplitude or to decline to 1/e of the maximal amplitude. The value for τ lies between a few ms to 50 ms, depending on the membrane considered. If the amplitude of the electrotonic potential is divided by the applied current, the membrane resistance R_m is obtained.

Since the time course of the electronic potential gives the kinetics of charging (or discharging) the capacitance across the membrane resistance, it follows that τ the membrane time constant, is equal to R_m times the membrane capacitance C_m. Therefore, knowing R_m and τ , it is an easy matter to get C_m. The value of the time constant is the same whatever the surface area of the membrane under consideration. Let us assume that we double the surface area of the membrane. The resistance will be half and the capacitance twice of the original surface. The time constant of this segment of membrane is thus:

$$\tau = 2C_m \frac{R_m}{2} = C_m R_m$$

Therefore it is clear that τ is solely defined by the properties of the membrane, it being independent of the shape or size of the cell.

If the recording electrode is inserted at various distances from the current electrode, the time course of the electrotonic potential becomes slower with increasing distance and the amplitude is also smaller. It can be shown as indicated in Fig. III.6.b that the amplitude Em falls exponentially with distance. The distance corresponding to a drop of E_m of 37 % (i.e. 1/e) is the space constant λ. It varies between 0.1 to 5 mm, depending on the cell.

This decline in E_m as the distance between current and recording electrodes increases is obviously related to the values of the longitudinal resistances (internal and external).

III.2.b. Electrotonic propagation

In electrical terms, every cell membrane is a leaky capacitance, i.e. a capacitance C_m in parallel with a resistance R_m, interposed between the external (R_e) and internal (R_i) resistances of the cytoplasm and the external fluid.

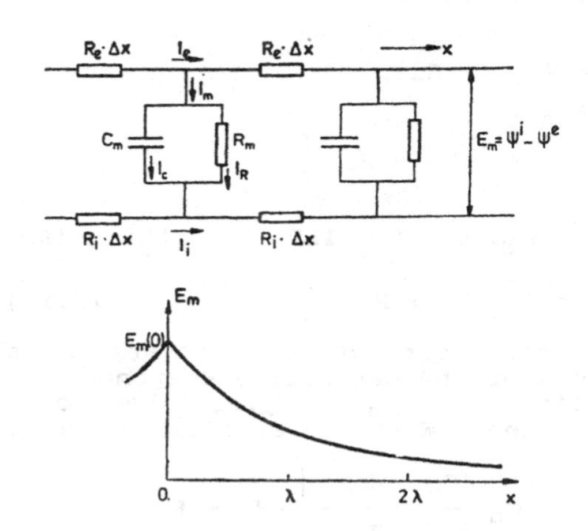

Figure III.6. a) Equivalent circuit representing the passive electric properties of elongated cells, particularly of the axon. The notations are Ie and Ii , the longitudinal external and internal currents; I_m *, transmembrane current consisting in a capacitive component (*I_C*) and a resistive one (*I_R*) ;* C_m *and* R_m *, membrane capacitance and resistance;* E_m *, the potential difference between the inside (*Ψ^i*) and the outside (*Ψ^e*).*

As it appears from (Fig. III.6) for long, quasi-cylindrical cells with the diameter d, such as the nerve and muscle fibres, these parameters refer to the unit of length (R_i and R_e), i.e. to a membrane area which is the unit of length times πd (R_m and C_m).

Ohm's law applied to extra- and intra-cellular media gives:

$$\frac{\partial \Psi}{\partial x} = R_e I_e \quad \text{and} \quad \frac{\partial \Psi}{\partial x} \; R_i I_i \qquad \text{(III.11)}$$

the variation along the distance of I_e and I_i being due to the transmembrane current I_m:

$$- \frac{\partial i_e}{\partial x} = I_m = \frac{\partial i_i}{\partial x} \qquad \text{(III.12)}$$

b) Exponential decay of a local membrane potential at different distances x from its origin.Taking into account that the transmembrane potential is $E_m = \Psi_i - \Psi_e$, from this and (III.11), one obtains:

$$\frac{\partial E_m}{\partial x} = R_i i_i - R_e i_e$$

and hence:

$$\frac{\partial^2 E_m}{\partial x^2} = R_i \frac{\partial i_i}{\partial x} - R_e \frac{\partial i_e}{\partial x}$$

On the basis of (III.12), this last relation becomes:

$$\frac{\partial^2 E_m}{\partial x^2} = I_m (R_i + R_e) \qquad \text{(III.13)}$$

The overall current I_m traversing the axonal membrane consists of the capacitive component: $i_c = C_m . \partial E_m/\partial t$ and the resistive one: $i_R = E_m/R_m$ (due to ionic flows through the membrane). With these, (III.13) reads:

$$\frac{\partial^2 E_m}{\partial x^2} = \left| C_m \frac{\partial E_m}{\partial t} + \frac{E_m}{R_m} \right| (R_i + R_e)$$

This second order equation with partial derivatives can be arranged in the form

$$- \lambda^2 \frac{\partial^2 E_m}{\partial x^2} = + \tau \frac{\partial E_m}{\partial t} + E_m = 0 \qquad \text{(III.14)}$$

where: $\lambda^2 = R_m/(R_i+R_e)$ and $\tau = R_m C_m$

The general solution depends on the initial and boundary conditions, a simple time independent solution being:

$$E_m(x) = E_m(0) \exp(-x/\lambda) \qquad \text{(III.15)}$$

The graph in Fig. III.6.b shows the exponential decrease of the electrotonic potential as the distance from the point of application increases. The length constant $\lambda[R_m/(R_i+R_e)]^{½}$ which characterizes the attenuation, can be expressed by means of parameters directly related to nerve fibre properties: the specific resistance per unit area of axonal membrane (r_m) and the resistivity of the axoplasm (r_i). If the diameter of the axon is d, then: $R_m = r_m/\pi d$ and $r_i = 4r_i/\pi d^2$.

Because of the much smaller value of the resistance of extracellular fluid (R_e) as compared to that of axoplasm (R_i):

$$\lambda \approx [R_m/R_i]^{\frac{1}{2}} = (r_m d/4r_i)^{\frac{1}{2}}$$

This relation permits the comparison of length constants of different fibres and thus of their ability to passively propagate potential changes. For example, an unmyelinated mammalian fibre with d = 1 μm has $\lambda \approx 330$ μm, for the large axons of crustaceans (d $\approx$ 100 μm, $\lambda \approx 2.5$ mm, and for the squid giant axon (d $\approx$ 500 μm), $\lambda \approx 5$ mm.

From physiological point of view, the main conclusion related to the passive propagation of electric signals in nerve fibres is that at a given point on the membrane the potential variations originating from various places arrive and summate. Thus, passive cable properties of the neurons represent the simplest biophysical basis of spatial summation, an essential aspect of integrative processes in the central nervous system. On the other hand, in view of the strong attenuation, quite pronounced even for large axons, the (passive) electrotonic spread is completely inadequate for long distance propagation of signals.

III.3. Regenerative propagation of action potentials in excitable membranes

The nature of the regenerative process by which the excitable membrane amplifies those depolarizations which exceed the threshold became somehow more apparent when Hodgkin and Huxley (1952a) showed the way in which the axonal permeability to sodium depends on transmembrane potential. At rest, membrane conductance for Na^+ is very low, but it steeply increases during depolarization. As a result, Na^+ ions enter the axoplasm, further increasing the depolarization, which in turn promotes sodium entrance. Thus, there is a self-sustained (regenerative) process which tends to continue until the inside of the fibre becomes positively charged at the value set by the electrochemical equilibrium potential of sodium, 50 mV. In fact this value is not attained because the increase in sodium conductance is a short transient process, followed by a slower and more persistent increase in potassium conductance. This makes potassium to exit from the axoplasm, a process which counteracts sodium entrance and brings the system to the resting value of membrane potential.

Hodgkin and Huxley (1952b) elaborated a detailed description of the variations in ionic conductances of the axonal membrane and implicitly of the ionic currents during the action potential.

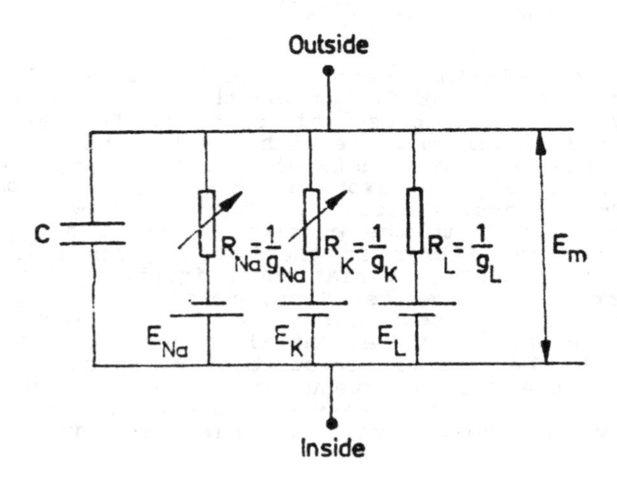

Figure III.7. Equivalent circuit of the giant axon membrane with the same notations as in Fig. III.6, and detailing the partial ionic resistances (R) and conductances (g). While sodium and potassium conductances R_{Na} and R_K are variable and depend on the membrane potential, the "leakage" conductance R_L is constant.

They used the equivalent circuit represented in Fig. III.7, which differs from that one in Fig. III.6 only because membrane conductances for Na^+ and K^+ (g_{Na} and g_K) are no longer constant, but depend on the instantaneous value of the membrane potential E_m.

Even since K.S. Cole and H.J. Curtis' early measurements of the giant axon impedance variations during the nerve impulse, it was observed that, while the resistance markedly decreases, the capacitance is practically unchanged (Cole, 1970). This is why in the equivalent circuit of the excitable membrane and in the equations describing the current, C_m is taken to be constant. However some studies (Takashima, 1976) pointed to the capacitance changes associated to intramembrane charge displacements when the ionic channels open and close.

Excitability of conducting cells is manifested by rapid transient changes of electrical properties. This is best exemplified by considering the intracellular electrical stimulation of the squid giant axon.

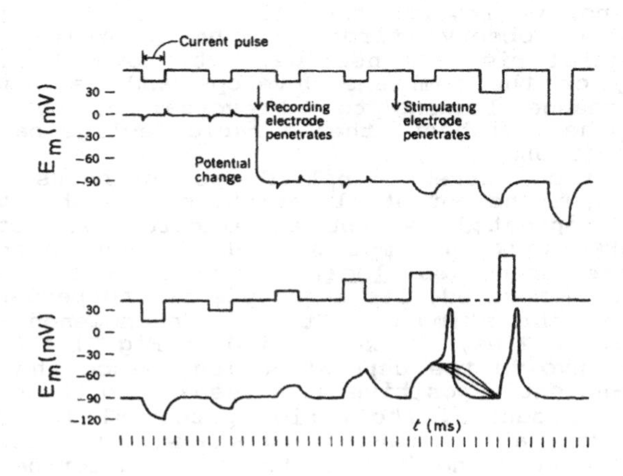

Figure III.8. Recording of resting and action potentials in an isolated giant axon. The recording electrode is close to the stimulating electrode thus explaining why the square pulse (stimulation) is distorted (after Katz, 1966).

Through the current electrode a square pulse of 2 ms duration is sent that hyperpolarizes the membrane.

The recording electrode, in close vicinity to the current injecting electrode, records the electrotonic potential which arises due to the cable properties of the axon. If one reverses the polarity of the stimulus, the square pulse depolarizes the membrane and for a given value, the threshold, the membrane potential jumps rapidly to a positive value and then returns more slowly to the resting potential, the whole variation being completed about 2 ms.

In Fig. III.8, it is seen that a depolarization of the membrne of 30 mV (thus bringing the membrane potential at - 60 mV) elicits the response. Moreover, the overshoot, i.e. the peak value of the action potential is around 120 mV. The rising phase of the action potential is not a mirror image of the descending phase or depolarization phase.

When the current applied is such that the membrane is hyperpolarized, we record a distorted signal because of the membrane capacitance. Increasing the amplitude of the square pulse, the recorded signal shows also a higher amplitude.

If now we reverse the polarity of the stimulating current, we do observe first, at the recording electrode the exponential rise and decline, but above threshold, an instability of the membrane develops and we observe the explosive change leading to a reversal of the membrane potential, the overshoot, then a rapid decline back to the resting conditions.

If the current applied as stimulus is near threshold, membrane potential variations are observed. They are never propagated, except as electrotonic potentials, and of course lack the amplitude of the action potential. This is the so-called local response. Once an action potential is generated, its magnitude is independent of the intensity of the stimulus. It is a phenomenon of short duration, about 2 ms, as exemplified in Fig. III.8 and does not merely involve the depolarization, since the axoplasm becomes transiently positive with respect to the outside.

Once produced, the action potential is propagated along the axon, at constant speed and without attenuation. It leaves the membrane in a state of refractoriness during which another action potential cannot be generated. This absolute refractory period has a duration of a few ms. Following it there is also a relative refractory period of longer duration and characterized by a higher threshold, thus meaning that a response can only be obtained if the intensity of the stimulus is increased. After these, the system is ready again to be stimulated and the action potential is propagated along the axon.

The ionic currents through the membrane are carried by sodium (I_{Na}), potassium (I_K) and also by other ions, providing a small and constant "leakage" current (I_L). Thus, the total membrane current, including the capacitive one, is:

$$I_m = C_m \frac{dE_m}{dt} + I_K + I_{Na} + I_L \qquad \text{(III.16)}$$

Under voltage-clamp conditions E_m is constant, so that the capacitive current vanishes and one can measure the ionic current alone. In short, the analysis of Hodgkin and Huxley proceeded as follows.

Clamping the membrane potential at a certain value E_m, one measures the time course of the current $I_m(t)$ which is in this case only the ionic current $I_i(t)$. Except for the leakage current, taken as a correction term, $I_m(t)$ consists of the components $I_K(t)$ and $I_{Na}(t)$. Their contributions can be separated if one replaces sodium in the external solution by the non-penetrating choline, in which case: $I_m(t) = I_K(t)$. It is assumed that each ionic flow through the membrane is driven only by its own electrochemical gradient (the "independence principle" of the ionic flows).

Fig. III.9 represents the separation within the total ionic current I_i of the two components I_K and I_{Na}. The leakage current is not shown, but its value can be known as the difference between I_i and I_K at the end of a long duration voltage clamp.

The ionic conductances are defined as the ratios of the corresponding ionic currents and the actual potential gradients, i.e. the differences between the electrochemical equilibrium potentials (E_K and E_{Na}) and the transmembrane potential (E_m):

$$g_{Na} = I_{Na}/(E_m - E_{Na}); \quad g_K = I_K/(E_m - E_K)$$

On the basis of a large set of experimental data, Hodgkin and Huxley concluded that the time course of membrane conductance for sodium can be best expressed as:

$$g_{Na}(t) = g^{\circ}_{Na} m^3(t) h(t) \quad \text{(III.17)}$$

where g°_{Na} is the maximum value attained and $m(t)$ and $h(t)$ are dimensionless parameters ranging from 0 to 1. This expression was not derived from a molecular description of processes involved in sodium permeation through the membrane, but merely chosen to provide the best fit of the experimental data which indicate a rapid activation and a slower inactivation of the conductance. Because the subunitary dimensionless parameters m and h can be viewed as the probabilities of some molecular events which occur inside the membrane and control sodium permeability, the above relation was interpreted as expressing the fact that Na^+ can pass through the membrane only when 3 "activating particles" open a channel and one "inactivating particle" does not block it.

The speculations about the detailed nature of

intramembrane events controlling the ionic passage prompted a certain reluctance in accepting the Hodgkin-Huxley theory.

The time variations of m and h parameters are represented by a direct process, with rate constants α_m and α_n, and an inverse one, with constants α_m and α_n, both assumed to be of first order (2)

$$\frac{d_m}{dt} = \alpha_m(1-m) - \beta_m \cdot m$$
$$\frac{d_h}{dt} = \alpha_h(1-h) - \beta_h \cdot h \qquad \text{(III.18)}$$

All rate constants depend on temperature, on Ca^{2+} concentration, and also on the membrane potential E_m. Hodgkin and Huxley expressed the results as:

$\alpha_m = 0.1(E_m + 25)/\{exp[(E_m + 25)/10]-1\}$; $\beta_m = 4\ exp(E_m/18)$ a.s.o.

these being just best fits of the empirical data.

Potassium conductance comprises only an activation process and it was represented by the relation:

$$g_K(t) = g^\circ_K . n^4(t) \qquad \text{(III.19)}$$

in which g°_K is the maximum value and n is again a dimensionless subunitary parameter. On the same line as before, this relation would imply that K^+ ions traverse the membrane when 4 simultaneous events with the probability n occur. The time dependence of n is given by:

$$dn/dt = \alpha_n(1-n) - \beta_n n$$

where α_n and β_n depend on the membrane potential in a similar way as it was exemplified above for Γ_m and β_m.

As mentioned before, the leakage current and thus the corresponding conductance g_L are constant.

If some potential changes propagate along the axon with the constant velocity θ, this represents the first time derivative of the distance reached by the wave front: θ = dx/dt, so that space derivatives appearing in cable

2 Significant deviations of the inactivation system h from first order kinetics were reported while some authors (Goldman, 1975), questioned even the concept of two separate processes governing the turn-on (m) and the turn-off (h) of the sodium current, others (Chin, 1977) accounted for the observed kinetics of the inactivation assuming it to be the second order process.

equations can be formed as:

$$\frac{d}{dx} = \frac{1}{(dx/dt)} \cdot \frac{d}{dt} = \frac{1}{8} \cdot \frac{d}{dt}$$

Now, from the relation (III.3) the whole (trans)membrane current is given by:

$$I_m = \frac{1}{\theta^2 (R_i + R_e)} \cdot \frac{d^2 E_m}{dt^2}$$

Replacing Im by the right hand in (III.16) and taking into account (III.17) and (III.19), one obtains for the membrane potential a second order differential equation which no longer contains any other independent variables:

$$\frac{1}{\theta^2 (R_i + R_e)} \times \frac{d^2 E_m}{dt^2} = C_m \frac{dE_m}{dt} + (E_m - E_k)\, \overset{0}{g}_k n^4 + (E_m - E_{Na}) \overset{0}{g}_{Na}\, m^3 h + (E_m - E_L) g_L \qquad \text{(III.20)}$$

Numerical solving of the above set of equations provided the time courses of the membrane potential (i.e., the action potential) and of membrane conductances for sodium and potassium as they are represented in Fig. III.9.

The equations (III.17)-(III.20) were established on the basis of voltage clamp measurements (that is unpropagated changes) and of the cable model of the giant axon. The fact that computed action potentials, resulting from the integration of these equations, are almost identical with real propagated action potentials in the giant axon of the squid, represented the strongest support for the usefulness of Hodgkin-Huxley theory.

Its successful application to the squid giant axon justified the efforts to devise similar treatments for other excitable systems, the most noteworthy being for the Ranvier node of myelinated fibres.

Frankenhaeuser and Huxley (1964), after analysing under voltage clamp conditions the ionic currents in *X. laevis* nodes, evolved a system of empirical equations of the same type, except for using membrane partial permeabilities P_i - which appear in the Goldman equation (III.8) - instead of the conductances g_i. Also, they had to consider an unspecific delayed component within the ionic current (I_p). Accordingly:

$$I_m = \frac{1}{\theta^2 (R_i + R_e)} \frac{d^2 E_m}{dt^2} = C_m \frac{dE_m}{dt} + I_{Na} + I_K + I_L + I_p$$

Goldman's equation expresses the ionic currents in the form:

$$I_i = P_i \frac{E_m F^2}{RT} \times \frac{C_i^e \exp(-FE_m/RT) - C_i^i}{\exp(-FE_m/RT) - 1}$$

with the lower index i standing for Na^+, K^+, L and p.

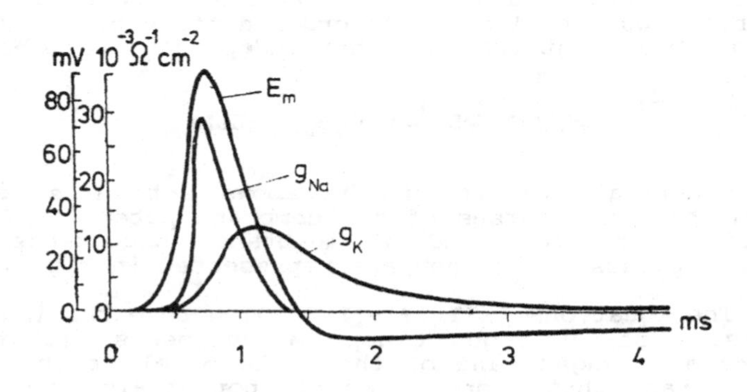

Figure III.9. Computed action potential and conductances obtained by numerical solving of Hodgkin-Huxley set of equations (after Hodgkin and Huxley, 1952).

The time variations of the permeabilities are given by the empirical equations:

$$P_{Na}(t) = P_{Na}^o\, m^2(t)\, h(t)\ ; \quad P_K(t) = P_K^o\, n^2(t)$$

$$P_p(t) = P_p^o\, p^2(t) \text{ and } P_L(t) = P_L^o$$

Again, the dimensionless subunitary parameters m, h, n and p are solutions of first order differential equations of the type:

$$\frac{dm}{dt} = \alpha_m (1-m) - \beta_m m$$

in which all rate constants α and β depend in a known way on Em, temperature and Ca^{2+} concentration in the bathing solution.

We conclude this section by stressing that the electric aspects of the nerve impulse are accurately described by the ionic theory, on the basis of which various equations of the Hodgkin-Huxley type were elaborated. Nevertheless, these equations are empirical and the theory provided only some suggestions about the molecular mechanisms of the processes involved in nerve impulse generation and propagation.

III.3.a. Ionic pathways in the axonal membranes

The above description of the action potential obviously implies that sodium enters the axoplasm from the outside (where it is more concentrated) and potassium leaves it. This was clearly proved since the early fifties with the use of radioactive isotopes ^{42}K and ^{24}Na, and it was estimated that on the whole duration (~2 ms) of an impulse in the giant axon, 1 cm^2 of membrane is traversed inwards by ≈ 3.10^{-12} mole Na^+ and outwards by an almost equal quantity of K^+.

The fact that voltage clamp data have shown the lack of any simple relation between the time onset of sodium and potassium currents was interpreted as indicating the passage of these ions through separate pathways in the membrane. The term "ionic channels "is of general use in the physiological-biophysical litterature, so that we also employ it, although we still consider that it is too suggestive of rigid structures and one of us (Schoffeniels, 1980; 1983) proposed the biochemically inspired term "ionic conductins" for the ion transporting proteins. The pioneering works of T. Narahashi and of B. Hille (in the sixties) have shown that 0.1 μM tetrodotoxin or saxitoxin (TTX, STX) in the bathing solution of a giant axon or a myelinated fibre produce the abolution of the sodium current (I_{Na}) while the other ionic currents (I_K and I_L) are not affected even by 100 times greater concentrations. TTX and STX are two hydrosoluble toxins extracted from the Japanese fish *Spheroides rubrides* and the dynoflagelates *Gonyaulax catanella* respectively.

When Na^+ is replaced by its possible substitutes (Li^+, NH_4^+) their currents are also abolished in the same manner, thus proving that the toxins do not have any specific interaction with sodium ions per se, but they block the channels through which sodium passes. When introduced internally into perfused giant axons, even 1 μM TTX or STX fails to block the sodium current; this indicates that the receptors of these toxins and consequently the molecular mechanism for closing sodium channels are localized on the outer side of the axonal membrane. On the basis of Hodgkin-Huxley equations, the inhibition of I_{Na} could in principle be attributed to either one of the following factors:

a) the decrease in the number of channels able to become permeable and thus of the maximum sodium conductance (Δg°_{Na}),

b) the decrease in the activation of sodium conductance(Δm) and ,

c) the increase in the inactivation of sodium conductance ($\Delta(1 - h)$). But voltage clamp experiments have shown that neither m nor h is affected by TTX, the only effect of the toxin being the diminution of the maximum sodium conductance i.e. the block of sodium channels.

Apart from these drugs which block the open sodium channel, there are two other types affecting other states of the channel: sea anemone toxin (ATX) and scorpion toxin (ScTX) prevent inactivation - i.e. affect the h system - while substances such as grayanotoxin (GTX), batrachotoxin (BTX) and veratridin stimulate the opening of the channel (an authoritative review of the neurotoxins as tools for pharmacological dissection of ionic channels in nerve membranes by Narahashi (1975) still keeps being informative).

The kinetics of TTX binding on the membrane indicates that it proceeds in 1:1 ratio with some receptors belonging to the molecular structures which represent the channels for sodium. Accordingly, the number of toxin molecules bound on the membrane gives the number of channels.

As TTX and STX act even in nanomolar concentrations, when a nerve is placed in a very small volume (≈50 μl) of solution having exactly that concentration which completely blocks I_{Na} (and thus impulse conduction) it will retain a quantity of toxin representing a significant fraction of the amount present in solution. If another nerve is then exposed to this solution, the degree of blocking I_{Na} can indicate - on the basis of the curve (Dose) / (inhibitory effect) - the amount of toxin which remained in solution, and thus that taken up by the first nerve. Apart of this bioassay method, the toxins can be made radioactive and the binding can be followed up with a radiation counter. The binding of radioactive labelled TTX indicated values rather different from those arising from the bioassay, and the binding of tritiated STX gave still different figures (Ritchie, *et al.*, 1976). Now the commonly accepted values of the number of sodium channels per μm^2 are: 110 (rabbit), 166-533 (squid) and 90 (lobster). On a theoretical basis, Hodgkin (1975) concluded that there is an optimum density of sodium channels at which conduction velocity is maximal, which for the squid axon would be just around 500/μm^2. Even keeping only the order of magnitude, it appears that the channels through which sodium ions passively enter the fibre are very sparsely distributed, they being several times less dense than the ($Na^+ + K^+$)ATPase through which sodium is actively pumped outside (≈750/μm^2 in the rabbit vagus nerve). The estimated number of sodium channels within the membrane of Ranvier nodes is much greater: 25.000/μm^2 in the rabbit sciatic nerve (Ritchie and Rogart, 1977) in accord with larger ionic current density in the nodal membrane.

The conductance of a sodium channel appears to be of the order of a few pS (= $10^{-12}\Omega^{-1}$), it being the ratio of the maximal conductance for sodium and the number of channels. Similarly, upon knowing the surface density of channels and the sodium flow, one estimates the number of ions passing through a channel. In the case of the squid giant axon, during the action potential ≈ 10^8 Na^+ ions enter the axoplasm in one second through one channel.

While TTX and STX selectively block sodium channels, there are other substances which act on potassium channels. It was known long ago that quaternary ammonium ions, especially tetraethylammonium (TEA), prolong the action potentials, induce repetitive firing in nerves and counteract the depolarizing effect of K^+ ions. TEA inhibits outward potassium current (I_K) without affecting the other components. For example, for frog Ranvier nodes 6 mM TEA decrease by 90 % I_K, without affecting at all I_{Na} and I_L.

TEA and TTX do not cross interact, for their effects are independent of one another, which clearly proves the functional independence of sodium and potassium channels in the axon membrane. The fact that 40 mM TEA produced a rapid and specific change of I_K when it is internally applied in perfused giant axons, while even 100 mM TEA did not induce any effect when externally applied, shows that TEA receptors are located on the inner face of the membrane. Another quaternary ammonium compound, the 4-aminopyridine, which was found to selectively block the potassium conductance in both myelinated and non-myelinated nerves, as well as in skeletal muscle fibres, acts equally from both the inside and the outside and blocks not only the outward K current, but also the inward one (Meves and Pichon, 1977). It is also remarkable that within the population of K^+ selective channels which contribute to the outward K^+ current, three distinct components can be separated pharmacologically on the basis of their sensitivity to TEA, 4-aminopyridine and to Ca^{2+} and Mg^{2+} ions (Thompson, 1977).

Armstrong (1975) estimated the density of potassium channels at about $10^3/\mu m^2$, but others have given lower values: 70/μm^2 in squid axon (Conti *et al.*, 1975) and 220/μm^2 in crayfish axons (Hucho, 1977). The conductance of a channel is 2-3 pS and the rate of potassium outflow through each channel during the action potential is 2-3 10^6 ions/s.

Similar results were obtained in a completely different way, from the study of "membrane electric noise", i.e. of the spontaneous fluctuations around the stationary values of the membrane potential or currents. In 1965, Verveen and Derksen reported random fluctuations of the resting potential in frog Ranvier nodes. Hyperpolarizations or depolarizations by a constant external current, as well as changes in the external potassium concentrations, all modified the intensity of these fluctuations; this suggested that they were involved in K^+ current through the membrane. An even stronger proof that electric noise in resting axon membranes arises from miroscopical fluctuations in the opening of potassium channels was that TEA markedly affects it, while TTX has no effect.

The alread mentioned fact that during the action potential the excitable membrane is crossed by ions only at discrete small patches of its area (the gateways or channels) together with the discrete nature of the charge carriers (the ions) obviously imply that what we are actually mesuring in the usual electrophysiological experiments are average values of some microscopical quantities. For instance, in voltage clamp experiments, the current recording technique averaged the individual currents flowing in a large assembly of channels. While from such macro-level values alone it is not possible to infer the characteristics of micro-level events, the posibility does exist if one records and analyses at an as high as possible sensitivity, the time fluctuations of either potential or current across the membrane. In the last twenty years the study of electric noise in nerve and muscle membrane became an usual tool complementing the macroscopical type of experiments (for a comprehensive review, see De Felice, 1981). The most useful means of characterizing noise (though not the only one) is the power density spectrum, obtained by a Fourier transformation of the potential or current fluctuations into the frequency domain. This spectrum immediately conveys how energy is distributed among the various sinusoidal components that make up the recorded fluctuation waveform.

As to its physical origins, the electrical noise present in excitable membranes arises from several sources (Stevens, 1972; Verween and De Felice, 1974):

a) the thermal agitation of the ions carrying the charge gives rise to a thermal (or Johnson-Nyquist) noise; from its spectrum, the passive impedance of the membrane may be inferred;

b) because the passage of each ion through the membrane is analogous to the movement of electrons from the cathode to the anode in an electron tube, there is also a shot noise similar to the shot effect in these tubes; its spectrum gives information about the average motion of individual ions within the membrane.

c) in systems with only a relative small number of available charge carriers it appears a flicker noise, also termed 1/f noise, because the amplitude of the power density spectrum is inversely proportional to the frequency; such noise was actually discovered in frog Ranvier nodes and in giant axons and its amplitude is connected with the number of charge carriers and with the mobilities of ions within the membrane.

Besides these electrical fluctuations, whose nature is purely physical and which are normally found even in non-biological system having ionic conductance, there is also a fourth one, more directly connected with the specific structure of excitable membrane. This is represented by conductance fluctuations arising from the random opening and closing of membrane channels or from their random transisitions between all the available conductance states. The analysis of these fluctuations make it possible to estimate the conductance of a single channel and then the number of channels per unit of surface area.

Though making inferences about membrane processes per unit of surface area is hindred by the difficulty in separating spectra from different sources, a wealth of results were obtained by appropriate physiological manoeuvers. For instance, near the resting potential the fluctuations are mainly connected with potassium conductance as they are strongly influenced by TEA but not by the specific blockers of sodium permeability such as TTX. From noise measurements the number of potassium gateways in node was estimated between 500 and $1000/\mu m^2$: later on the components depending on sodium permeability were identified and separated into the distinct contributions of the activation and inactivation mechanism (the $\underline{m}$ and $\underline{h}$ parameters) (Conti *et al.*, 1976). Even some old empirical facts, such as the existence of critical slopes of stimuli, received a microscopical significance (Zaciu and Margineanu, 1979).

When analysing the dependence of sodium and potassium conductance g_{Na} and g_K on the membrane potential, Hodgkin and Huxley interpreted the permeability changes for these ions as resulting from the electric field induced reorientation of some charged or dipolar membrane components. The conductance gNa would be controlled by negatively charged molecules which, depending on the membrane potential E_m, are in a Boltzmann distribution between the inside and the outside of the membrane, the conductance being proportional to the number of inside oriented molecules. To explain the steep dependence of g_{Na} on E_m, they assumed that each channel is controlled by 6 electronic charges and concluded that a small outward current would precede I_{Na}, but it is completely masked by this. Their theoretical conclusion was confirmed two decades later, when these "gating" currents, associated with intramembrane charge movements could be experimentally detected (Armstrong and Bezanilla, 1973).

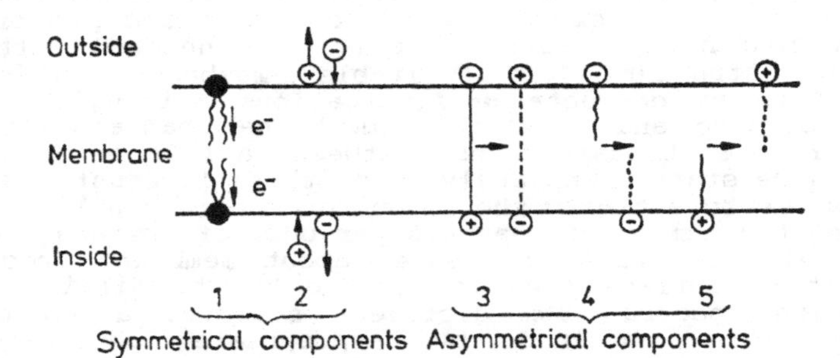

Figure III.10. Different origins and components of the capacitive current in the axonal membrane in response to a depolarizing pulse.

To measure the currents associated with the inward displacement of the negative charges controlling sodium channels, Armstrong and Bezanilla (1973; 1974), then Keynes and Rojas (1974) blocked both I_{Na} (the bathing solution was sodium free and contained TTX) and I_K (the giant axons were internally perfused with CsF). The capacitive current has some symmetrical components which express electronic redistributions on the apolar molecules and ionic displacements between the surface of the membrane and the electrodes (Fig. III. 10).

These components are identical for both depolarizing and hyperpolarizing pulses but the ionic channels open only for depolarizing pulses so that the gating current is asymmetrical. To substract the symmetrical components, 10 to 100 depolarizing, and then an equal number of hyperpolarizing pulses were applied, and the resulting capacitive currents were added.

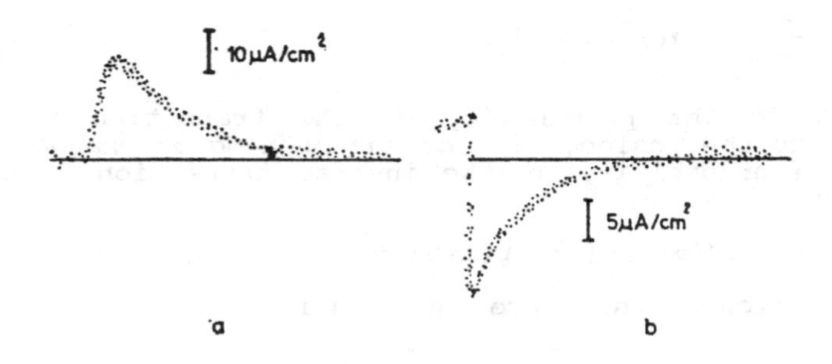

Figure III 11 Gating currents at (a) the beginning and (b) the end of 70 mv depolarizing pulses (form Armstrong and bezanilla, 1974)

The time course of gating currents (Fig. III.11) can be accounted for starting from the above mentioned hypothesis of a system of N charged particles able to independently undergo first order transitions between the inside and the outside of the membrane. If at a certain value V of the transmembrane potential, N_1 particles are inside and $N_2 = N - N_1$ outside, their ratio, given by Boltzmann distribution is:

$$(N_1/N_2) = \exp\,[-z'e(V - V')/kT]$$

where z′ is the actual valency of each particle, e is the electronic charge, k is Boltzmann's constant, T is the absolute temperatue and V′ is the value of membrane potential for which the particles are uniformly distributed between the inside and the outside: $N_1 = N_2$. Inside there is a fraction:

$$p = N_1/(N_1 + N_2)$$
$$= \exp[-z'e(V - V')/kT]/(1 + \exp[-z'e(V - V')/kT])$$

of the total number of particles.

Upon assuming first order transitions, the variation of this fraction p is:

$$\frac{dp}{dt} = \alpha(1-p) - \beta \cdot p$$

where α is the probability of the transition from the outside to the inside, i.e. of the opening of channels, and β is the probability of the inverse transition. With the notations:

$$\tau = 1/(\alpha + \beta) \;; \quad p(t=0) = p_o \;; \quad p(t\text{-->}\infty) = p_v$$

the solution of the above equation is:

$$p = p_v + (p_o - p_v) \cdot \exp(-t/\tau)$$

The whole electric charge of the assembly of particles is:

$Q_{tot} = z'e(N_1 + N_2)$, so that the gating current is:

$$i = Q_{tot}\frac{dp}{dt} = Q_{tot}\left(\frac{p_v - p_o}{\tau}\right) \cdot \exp(-t/\tau)$$

The above equation accounts for the exponential decay of the asymmetric intramembrane currents, the time constant τ depending on both the amplitude of the pulse and the temperature.

It is worth mentioning that the whole charge transferred at the onset of a pulse from one face to the other of the membrane is equal with that moving oppositely at the end of the pulse and also that temperature does not influence the magnitude of this total charge. These prove that the charges carrying the asymmetric currents actually are parts of the membrane itself. Assuming that each sodium gateway would be controlled by six elementary charges, the total charge displaced in squid axon membrane indicates a number of several hundred gateways per μm2, in good agreement with TTX binding data and with the values derived from electrical noise measurements.

Following their first measurements fifteen years ago, the gating currents have been detected in all excitable cells where they have been looked for (for review see Almers, 1978).

While the mere occurrence of differences in capacitive transients during opposite pulses is easily understandable in view of the structural asymmetry of biological membranes, their physiological role is not so easy to assess and the conclusions concerning the positioning within the channel and the effective valence of the gating particles are still subject of debate (Bezanilla *et al.*, 1982; Keynes, 1986; Rayner and Starkus, 1989). But, within the context of the phenomenological aspects of bioelectricity we do not pursue further these aspects, to which some reference will be made when approaching the molecular structure and transitions of ionic gateway proteins

III.4. Intercellular transmission of excitation

The nervous system is a network of excitable cells which carry to and from the different parts of a complex animal organism the data necessary to the coordination of their functioning and the integration of that animal in its environment. For performing this function, the signals, i.e. the action potentials have to be orderly transmitted from one excitable cell to other(s), a need which imposed the existence of specialized cellular structures - the synapses - at the contacts between neurons and between these and the effector muscle and gland cells. The synapses ensure the structuring of the (sometimes enormous number of) neurons into the finest hardware one can imaginate - the nerve centres and networks (Steriade and Llinas, 1988) and are the first support of the cellular processes involved in higher cognition (Byrne, 1987). The neuromuscular junctions bear the special name of end plates (EP).

The transmission of electrical excitation from one cell to the other is often mediated by the secretion and release from the first cell of a chemical transmitter, which binds to specialized receptor proteins on the membrane of the second one and induces the appearance of the action potential into it. In some rather particular cases, the neurons communicate directly by means of local currents. These two possibilities are respectively associated with <u>chemical</u> and with <u>electrotonic synapses</u>. There are several major structural differences between these (Kandel and Siegelbaum, 1985):

a) while in chemical synapses the contacting area

between cells is very small when compared with the total surface of the post-synaptic cell, in electronic synapses it is significant;

b) the space between the membranes of the two cells (the synaptic cleft) is approximately 200 Å or greater at chemical synapses while in the electrotonic synapses a narrow gap of only about 20 Å width exists so that the membranes of the two cells appear almost merged.

The electrotonic junctions between all excitable and non-excitable cells possess a particular type of structure called gap junctions. These are intercellular communicating channels which allow maximum current flow between cells, with little or no current leakage into the extracellular space; they constitute the ideal anatomical structure for electrotonic intercellular communication (Hertzberg *et al.*, 1981).

c) at the chemical synapses, the pre- and postsynaptic regions have different ultrastructures, while a remarkable symmetry is seen in electrotonic synapses. Thus, at chemical synapses numerous minute bodies, approximately spherical, are found in the presynaptic zone, which is also rather rich in mitochondria (Gershon *et al.*, 1985).

On the basis of cable properties, several physiologists, particularly Katz (1966) and Bennett (1977) have discussed the specific structural requirements for electrotonic transmission. Thus, according to equation III.15, the passive propagation makes a potential variation to decrease exponentially with distance.

As the excitation threshold is about 20 mV and the amplitude of the AP is around 100 mV, it follows that even if a certain zone of membrane is rendered non-excitable (by local cooling or by anaesthesia) the signal can be electrotonically propagated over a distance x given by the condition:

$$\exp(-x/\lambda) \geq \frac{20}{100} \quad \text{, or} \quad x \leq \lambda \ln 5$$

The spatial constant λ (defined in §III.3) is determined by the electrical properties and the diameter of the axon. In a unmyelinated axon 5 μm in diameter, $\lambda \approx 1$ mm and therefore the impulse can electrotonically propagate over a length of at least 2 mm. But this propagation will be stopped if the cable structure is no longer continuous and a break occurs. The impedance of the cable equivalent to that axon is 2.10^7 Ω (Fig. III. 12a). If the axon is sectioned and the two ends are covered with the same type of membrane and separated by a 150 Å space, filled with interstitial fluid, a signal coming from the first portion will meet an input impedance of 3.10^9 Ω (corresponding to the pre-synaptic membrane), shunted by a resistance of 3.10^6 Ω corresponding to the synaptic cleft and being continued by 3.10^9 Ω (the post-synaptic membrane) in series with the 2.10^7 Ω impedance of the cable (Fig. III.12b).

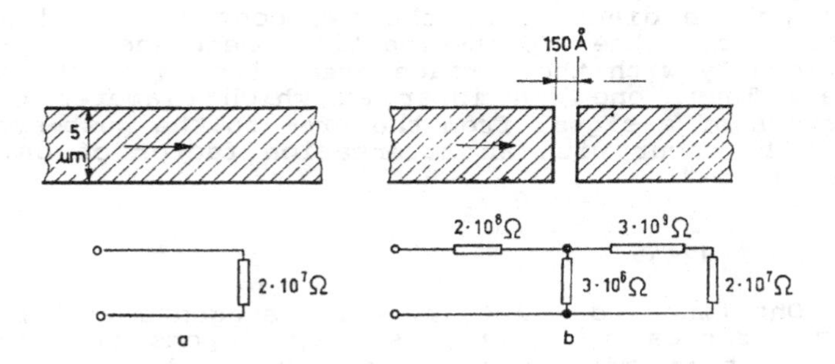

Figure III.12. Equivalent impedances for the transmission of the action potential: (a) along a non-myelinated axon with the diameter of 5µm and (b) at the junction between two axons separated by a space of 150 Å with interstitial fluid (from Katz, 1966).

Even when the pre-synaptic membrane is in the active (excited) state, its impedance being reduced to 2.10^8 Ω instead of 3.10^9 Ω, an electric signal will not pass from the pre-synaptic portion to the post-synaptic one, because there is an attenuation coefficient of 10^{-4}, so that it is impossible that the excitation threshold would be reached. Indeed, the signal propagating in the "post-synaptic cable" (of impedance 2.10^7 Ω) is proportional to the ratio $2.10^7/3.10^9$, the difference being attenuated by the post-synaptic membrane. On the other hand, into the post-synaptic zone only a fraction of $3.10^6/2.10^8$ of the initial signal penetrates, the rest passing through the shunt of the synaptic cleft. Thus, the attentuation coefficient of the electric signal is

$(2.10^7/3.10^9) \times (3.10^6/2.10^8) = 10^{-4}$. In neuromuscular junctions, the electric attenuation is even greater, since the synaptic cleft is larger than 150 Å, and the surfaces of the axon terminals are much smaller than that of the muscle fibre; therefore they cannot provide the fibre with sufficient current to depolarize it. It is clear that in such structural conditions, the propagation of an excitatory electric signal from the pre- to the post-synaptic region is impossible.

However the situation might become different as the area of contact increases. While the shunt resistance of the synaptic cleft depends only on the width of the cleft and not on the diameter of the two contacting cells, the impedance of the post-synaptic membrane decreases proportionally with the surface area, i.e. with the square diameter. Thus, one can infer at which diameter *d* the propagation of a signal from the pre- to the post-synaptic region will occur with an attenuation factor of only 1/5 instead of 10^{-4}:

$$10^{-4}(d/5)^2 = 1/5$$

One finds: d = 224 μm, which suggests that between the giant fibres of arthropods, cephalopods and mollusca the propagation of the electric signal at intercellular junctions is quite possible.

The conclusion is that the action potential can sometimes be directly transmitted between excitable cells on the basis of electrical properties of membranes alone, but most often it is propagated through chemical mediators which transiently modify the electrical properties of the post-synaptic membrane. The main functional characteristics of the chemical synapses is that excitation propagates unidirectionally, so that they act as valves, allowing only orthodromic impulse propagation in multineuronal chains. Between the arrival of the action potential at pre-synaptic axon terminals and the depolarization of the post-synaptic membrane, a latency of about 0.5 ms occurs. This time lag is required by the chemical mediation. On the contrary, in electrotonic synapses, the excitation propagates bidirectionally, though one of the directions could be preferential, and it spreads from one cell to the other with the same velocity as the conduction of action potential in the nerve fibre, without any delay. The synaptic transmission is achieved at least in higher vertebrates via chemical mediators. This is why we shall focus on the phenomenology and then on the molecular mechanisms of this type of transmission, with particular emphasis on the end plates which is the major example.

III.4.a. Main events in chemical synaptic transmission

Having in mind the essential structural features of the intercellular synaptic contacts of the kind schematized in Fig. III.13, one can understand what happens when one (or a series of) action potential(s) propagates up to the axon terminals of the motorneurone.

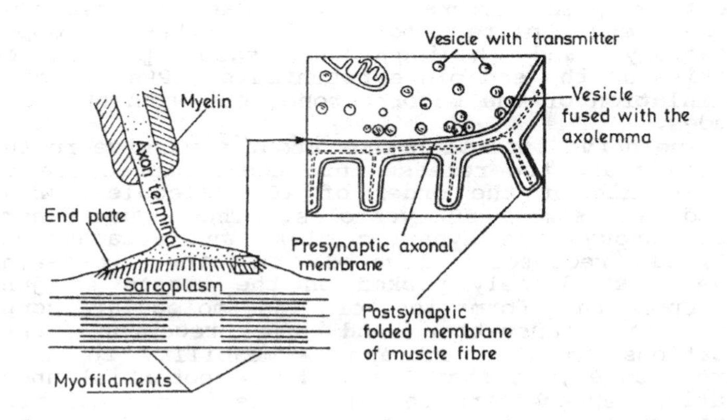

Figure III.13. Schematic representation of the structure of neuromuscular junctions.

In the perikarion of the presynaptic neuron there is a continuous synthesis of that substance which acts as neurotransmitter - which in the case of motorneurons is acetylcholine. It is transported as vesicles and stored in the axon terminals. The role of the quasi-identical vesicles, which were revealed long ago by electron micrographes and their very existence as normal cellular components is subject of debate, but this topic is of lesser significance for the purpose of our presentation.

The chemical nature of the substances secreted by neurons, by which synaptic transmission is achieved, distinguishes this general type of neurosecretion from the more specialized one, in which the product is a hormone, as for example the hypophyseal hormones. Nevertheless, the mechanism of transmitter secretion by all neurons, of hormones by the neurosecretory ones and even the mechanisms of secretion from non-neural cells are rather alike. They involve the storing of the secretion product in the form of granules or vesicles and their release when the prejunctional membrane is depolarized by the arriving action potentials.

In the resting state, some of the transmitter vesicles spontaneously attach to the post-synaptic membrane, fuse with it and discharge their contents into the synaptic cleft. The transmitter diffuses through the synaptic cleft and induces small local depolarizations of the post-synaptic membrane which can be recorded as miniature end plate potentials (mEPPs) appearing spontaneously and having the same pharmacological sensitivies as the end plate potentials (EPPs) elicited by the stimulation of the motorneurone, but with much smaller amplitudes.

The nerve impulse brings about a massive rupture of the vesicles and the release into the synaptic cleft of a qantity of ACh of the order of 10^7 molecules, which is contained in some 200 vesicles. The neurotransmitter diffuses through the synaptic cleft and attaches to the specialized receptors located on the post-synaptic membrane, most densely packed on the crests of junction folds. Upon the formation of the molecular complexes between the transmitter and the receptor, specific modifications in the membrane permeability to different ions are induced, so that the membrane potential undergoes a variation representing the EPP. This is a local response too, but, when reaching the threshold value of the specific post-synaptic cell, it generates the propagated AP, which in muscle fibres triggers the contraction. On the post-synaptic membrane there is an enzyme (the acetylcholinesterase - AChE) which hydrolyzes the transmitter (ACh) spliting it into precursors (choline and acetate). These are partly reabsorbed in the presynaptic region, and used for the subsequent transmitter synthesis.

The postsynaptic potentials (PSPs) might rise or fall in amplitude, depending on the history of presynaptic activity. These changes are the first electrophysiological manifestation of the dynamics of neural functions. At some synapses PSPs, elicited by a repetitive stimulation, might rise to many times the size of an isolated PSP, the growth occuring in less than 1s and decaying equally fast. This is the synaptic facilitation. If PSP amplitude gradually rises during tens of seconds of stimulation, the phenomenon is post-tetanic potentiation. The enhanced synaptic transmission over an intermediate timescale of a few seconds is called augmentation. These manifestations of short-term plasticity in synaptic transmission (Zucker,

1989) can be attributed to the accumulation of Ca^{2+} and Na^{+} in the nerve terminal after repeated action potentials. The increased internal sodium concentration further elevates internal Ca^{2+} which causes the potentiated transmitter release.

Because the chemical mediator is stored in the presynaptic endings as quasi-identical vesicles which, when they rupture, discharge all their contents into the synaptic cleft, it is expected that the EPP induced by nerve stimulation should have a quantal character, that is its amplitude should be a whole multiple of the depolarization produced by a "quantum of chemical mediator", i.e. by the contents of a vesicle.

The classical studies by del Castillo and Katz and by Boyd and Martin in the fifties established the quantal character of chemical transmitter release, as expressed by the quantized nature of EPP. The conclusion of these studies, are summarized in the following statements (Martin, 1977):

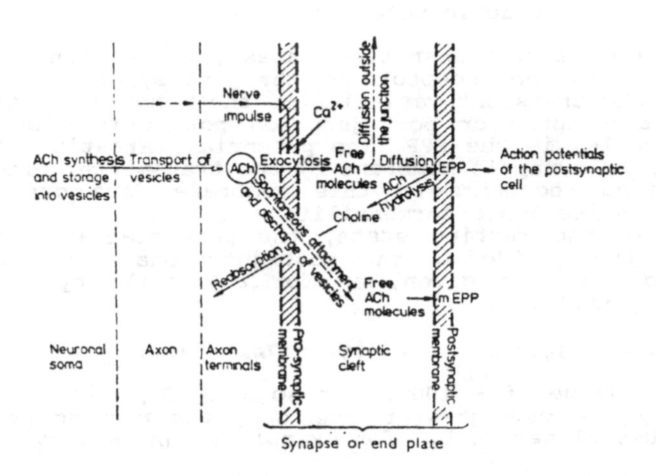

Figure III.14. Synopsis of main processes involved in chemical synaptic transmission.

1) The neurosecretion of chemical mediators occurs in the form of quanta, it resulting from the discharge into the synaptic cleft of the whole contents of those vesicles

which are ruptured;

2) In the resting state of the pre-synaptic nerve terminals, the rupture of vesicles is a spontaneous and random phenomenon with a very small probability of occurrence, as it is shown by the mean frequency of only one per second;

3) mEPPs represents the electrical events produced in the postsynaptic membrane by this spontaneous neurosecretion; therefore they also have a random and quantal character;

4) The nerve impulses arriving at the pre-synaptic endings greatly increase the probability of rupture of the vesicles and the number of those which simultaneously rupture, but do not modify the contents of each of them (for a more recent review see Kandel, 1985).

On the basis of these ideas, the main events involved in the chemical transmission of excitation between two cells, particularly a motorneuron and a muscle fibre innervated by it, can be schematically represented as in Fig. III.14.

III.4.b. Post-synaptic potentials

The neuro-transmitter causes, as a result of its attachment to the receptors on the post-synaptic membrane, a local and transient variation of the membrane potential - the post-synaptic or post-junction potential - which, in muscle cells, is the EPP. The potential variations induced in the post-synaptic membrane by the transmitter binding to specialized receptors on this membrane, are due to the changes in its ionic permeabilities.

In the resting state, the potential difference E between the inside of the post-junctional cell and the synaptic cleft is given, as for any cell, by Goldman's formula (III.4):

$$E_m = (g_K E_k + g_{Na} E_{Na} + g_{Cl})/(g_K + g_{Na} + g_{Cl})$$

E_K has a value of - 100 mV or so and E_{Na} is about + 50 mV. As g_K is much greater than g_{Na}, the resting potential has values closer to E_K, for example around -70 mV.

If the effect of the chemical transmitter is to open the channels through which a certain type of ion passes, it will make the membrane potential E_m to approach the electrochemical equilibrium potential of the respective ions. When the membrane becomes permeable to Na^+ ions, the potential variation will mean a reduction in the negative value of E_m , thus a depolarization. In this case, the post-synaptic potential is excitatory and, when reaching the threshold value, it will elicit an action potential in the post-synaptic cell. In the particular case of the end plate, EPP is always excitatory. But in those cases when the transmitter increases the permeability of the membrane to K^+, it causes a hyperpolarization and the post-synaptic potential will be inhibitory.

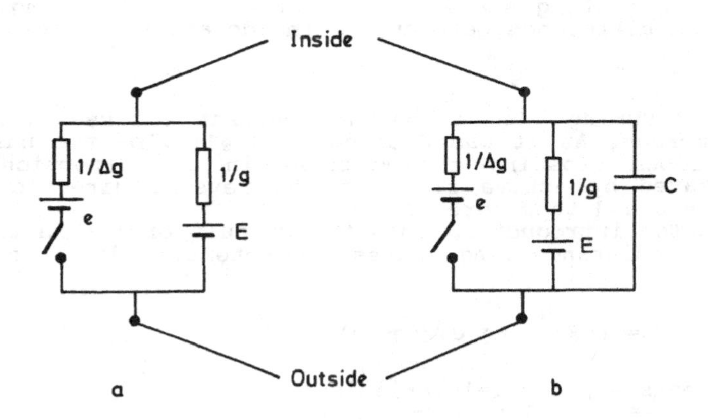

Figure III.15. The equivalent circuit of a post-synaptic membrane, in the resting state (a) and in the presence of the transmitter (b).

If in the post-synaptic membrane the transmitter opens poor by selective channels through which several ionic species are able to pass, the membrane potential will not be identical with the electrochemical equilibrium potential of any ion and it is termed equilibrium potential of the transmitter. It is the most important experimental parameter characterizing the modifications produced by the transmitter in the post-synaptic membrane. In striated muscles, ACh brings about an increase in the conductance of the post-synaptic membrane for both Na^+ and K^+ (and even for other ions), in such proportions that the equilibrium potential of the transmitter is excitatory, with values around - 20 mV.

The form and amplitude of the conductance variation Δg, elicited in the post-synaptic membrane by the transmitter, are explained by the equivalent circuit in Fig. III.15a.

The battery E_m represents the resting transmembrane potential, Δg is the total conductance of the resting membrane and the transmitter effect to open new ionic channels is represented by the switch which introduces in parallel into the circuit the conductance δg, together with a battery whose value <u>e</u> is that of the equilibrium potential of the transmitter (Takeuchi, 1977; Kandel and Siegelman, 1985).

In resting state, in the absence of e and Δg, the potential difference between the inside and the outside is:

$$V = E_m + i/g,$$

assuming that a constant current i passes through the membrane. As it was discussed in §III.1b, the passive ionic flows crossing the membrane in one direction are compensated by active flows in the reverse direction, so that $i = 0$ and therefore $V_o = E_m$.

The introduction into the circuit of the battery e and the resistance $1/\Delta g$ makes the potential difference to become:

$$V = (gE_m + \Delta g\ e)/(g+\Delta g)$$

which means a potential variation:

$$\Delta V = V - V0 = \Delta g(e-E_m)/(g+\Delta g)$$

with respect to the resting state.

The circuit in Fig. III.15b can also account for the time variation of the post-synaptic potential. The presence of the membrane capacity makes the potential variations not to be instantaneous, but with one rising and one exponentially falling phase. If the action of the chemical mediator is to open briefly the ionic channels upon the introduction into the circuit of the battery e and conductance δg for a time T, then the rising phase of the potential variation will last at most until the moment T. The instantaneous value of δV is:

$$\Delta X(t) = \frac{\Delta g(e-E_m)}{g+\Delta g}[1-\exp(-t/\tau)] \qquad \text{(when } t \le T\text{)}$$

Here τ is a time constant (see below). When the switch is reopened, the maximal reached value, $\Delta V(T)$, will fall exponentially with another time constant τ':

$$\Delta V(t) = \frac{\Delta g(e-E_m)}{g+\Delta g}[1-\exp(-T/\tau)] \exp(-t/\tau') \quad \text{(for } t>T$$

The time constants τ and τ' respectively correspond to the circuits with and without the conductance δg, so that:

$$\tau = C/(g + \Delta g) \quad \text{and} \quad \tau' = C/g$$

If the amplitude of the post-synaptic potential exceeds the value of the excitation threshold, an action potential will be elicited in the post-synaptic cell. When the local post-synaptic potential does not exceed the threshold, or it is inhibitory, it will add to other local potential variations, arising from other nerve terminals, the total effect depending on the resulting amplitude. Because of the parallel insertion of e and $1/\Delta g$ and of C, the circuit in Fig. III.15b explains the fact that the post-synaptic potentials are algebrically summed both in time and space, this phenomenon representing the very basis of the integrative functions of synapses. The spatial and temporal summation is much more exquisitely performed at the chemical synapses than on the mere basis of passive cable properties.

Summing up, the representation of the post-synaptic membrane by means of equivalent circuits of the type shown in Fig. III.15 allows describing the junction potentials produced by synaptic transmitters, within the same conceptual framework of the ionic theory as for the propagation of the action potential within a single excitable cell. The understanding we get from such descriptions remains however as limited as is the ionic theory itself and had to be implemented by molecular approaches. An important intermediate step between the electrophysiological measurements and the molecular descriptions of the synaptic events was the study of the transmitter noise in the post-synaptic membranes, initiated by Katz and Miledi (1972).

In the vicinity of the end plate an intracellular microelectrode records not only the bioelectric events of the type of mEPP but also a noise which is not simply produced by the various physical sources (minute mechanical shiftings of the electrode, the electronic <u>noise</u> of the equipment). When the transmitter (ACh) is added in the bathing solutions, at concentrations low enough not to induce excitation, it causes a constant depolarization, but also an increase of the noise, owing to the fluctuations of the channels opened by ACh. This additional noise does not occur when similar depolarizations are produced by other means than ACh and does not depend on the method of application of acetylcholine. This clearly indicates that it is associated with the molecular interaction between the chemical transmitter and the membrane.

The electrical noise of the post-junctional membrane can be analysed to establish the amplitude and distribution of the elementary events. On the basis of these parameters, Katz and Miledi (1972) calculated that in the frog muscle the ionic channel opened by ACh has a conductance of 10 pS; it allows the passage of a current of 10 μA in 1 ms, that is 5.10^4 monovalent cations.

The depolarization ΔV brought about by ACh results from the summation of identical elementary effects, each of them occurring at random time intervals, with a mean frequency n and having the same time evolution given by a certain function f(t). With the quite plausible hypothesis that the elementary potential variation has the same form as the mEPP, but a much smaller amplitude and a greater rapidity, it may be considered to consist of an instantaneous increase until the amplitude a is reached, followed by an exponential fall with a certain time constant τ:

$$f(t) = a \cdot \exp(-t/\tau)$$

Thus a is the amplitude of the local variation of potential, due to the opening of a single ionic channel. ΔV is the sum of all elementary effects, i.e.:

$$\Delta V = n \int_0^\infty f(t) \cdot dt = n \int_0^\infty a \cdot \exp(-t/\tau)dt = n \cdot a \cdot \tau$$

The noise amplitude is the mean square deviation of the depolarization:

$$\overline{E^2} = n \int_0^\infty f^2(t) \cdot dt = \tfrac{1}{2} n a^2 \tau$$

It follows that the parameter a can be calculated upon measuring the depolarization ΔV produced by the given ACh concentration and by the noise amplitude

$\overline{E^2}$: $a = \overline{E^2}/\overline{\Delta V}$.

In the original experiments of Katz and Miledi (1972) it was found that a ≈ 0.3 μV.

The 10^3 ratio between mEPP amplitude and the local

variation of potential produced by the opening of a single channel, suggests that a quantum of transmitter opens 1000 ionic channels.

III.5. An overview of bioelectric phenomena

The rapid flow of information within the excitable cells, between them and even from them to some non-excitable cells is conveyed by electrical signals. These are transient changes of the membrane potential (Em), which is determined primarily by the intra- and extracellular concentrations of K^+, Cl^- and Na^+ and by membrane permeabilities toward these ions. In the resting state, Em is close to the Nernst potential of K^+, to which the membrane is twice more permeable than for Cl^- and more than 20 times than for Na^+. While the Cl^- ions might be often considered at equilibrium between the cytoplasm and the outside, both K^+ and Na^+ are not. There are a steady efflux of K^+ out of the cell and a steady influx of Na^+, both passing through channels which act as dissipators of the existing electrochemical gradients and which are non-gated. This means that these channels are passive, in the sense that their conductance properties are not changed by stimuli. The passive (entropic) flows of Na^+ and K^+ are balanced by the Na-K pumps, which maintain the ionic gradients at the expense of metabolic energy (in the form of the ATP which they hydrolize).

The electrical signals are the action potentials, the synaptic potentials and also the receptor potentials, these later ones being elicited by all kind of stimuli in the receptor cells (for reviews see chapter VI in the monograph by Vasilescu and Margineanu, 1982) or the chapter 23 by Martin (1985) in the "Principles of Neuroscience" edited by E. Kandel and J. Schwartz).

The changes in E_m that occur during all kinds of signals are caused by important changes in the membrane permeabilities to the three main ions, brought about by the gating of a special set of channels whose opening is triggered by either electric field or by ligand molecules. In particular, the action potential is generated by the flow of ions through voltage-gated Na^+ and K^+ channels (Fig. III.16).

Na^+ and K^+ move through two-distinct and independent populations of channels, both being sparsely distributed membrane proteins which open and close in an all-or-none fashion when their gating charged groups are rearranged by voltage changes. Apart these channels, specific for the conducting membranes (e.g. the axonal membrane) there are other important membrane channels such as the Ca^{2+} channels in the pre-synaptic membranes and the chemically gated channels in the post-synaptic membranes.

All the bioelectric events are usefully modelled by equivalent electric circuits (for a classical account see Cole, 1970 and for a more recent presentation Koester, 1985) which include conductive (dissipative), capacitive (reactive) and electromotive force (active) components. These are to be respectively attributed to ionic channel proteins, to the lipid bilayer of the membrane and to the ionic pumps.

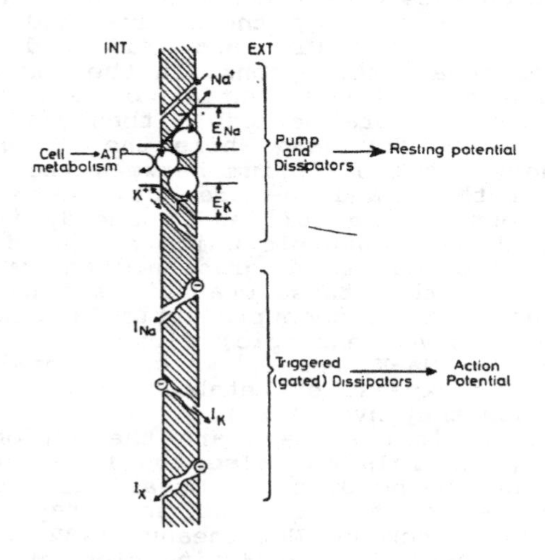

Figure III.16. Thermodynamically-inspired representation on the monoequilibrium distribution of microions across a plasma menbrane. The dowhill flows through the dissipators (channels) are balanced by the uphill flows, driven by the pump. This consumes ATP and maintains unveven ionic distributions. The turning wheels representations of the ionic pumps have no reality, but are aimed to suggest the couplings and the possibility of reversal. In the excitable membranes, a separate set of channels are triggered by the stimuli and they transienly dissipate part of the electrochemical gradients created by the pumps (from Margineanu, 1987)

References

Almers, W. (1978) Rev. Physiol. Pharmacol. 82, 6-190.
Armstrong, C.M. (1975) Quart. Rev. Biophys. 7, 179-210.
Armstrong, C.M. and Bezanilla, F. (1973). Nature 242, 459-461.
Armstrong, C.M. and Bezanilla, F. (1974). J. Gen. Physiol. 63, 533-552.
Bennett, M.V.L. (1977) pp. 357-416. In "Handbook of Physiology", vol. I, part I, E.R. Kandel (ed.) American Physiol. Soc. Bethesda, MD.
Bezanilla, F., Taylor, R.E. and Fernandez, J.M. (1982) J. Gen. Physiol. 79, 21-40.
Byrne, J.H. (1987). Physiol. Rev. 67, 329-439.
Chiu, S.Y. (1977) J. Physiol. 273, 573-596.
Civan, M.M. (1978) Am. J. Physiol. 234, F261-F269.
Cole, K.S. (1968) Membranes, Ions and Impulses. Univ. of California Press, Berkeley.
Conti, F., Hille, B., Neumke, B., Nonner, W. and Stèmpfli, R. (1976) J. Physiol. 262, 699-727.
DeFelice, L.J. (1981) Introduction to Membrane Noise, Plenum, New York.
Frankenhaeuser, B. and Huxley, A.F. (1964) J. Physiol. 171, 302-315.
Gershon, M.D., Schwartz, J.H. and Kandel, E.R. (1985) Ch. 12 in "Principles of Neural Science" (2nd edition) E.R. Kandel and J.H. Schwartz (eds.) Elevevier, New York.
Goldman, L. (1975) Biophys. J. 15, 119-136.
Heinz, E. (1981) Electrical Potentials in Biological Membrane Transport (Ch. 1) Springer, Berlin.
Hertzberg, E.L., Lawrence, T.L. and Gilula, N.B. (1981). Ann. Rev. Physiol. 43, 479-491.
Hille, B. (1984) Ionic Channels of Excitable Membranes, Ch. 9, Linauer, Sunderland, Mass.
Hodgkin, A.L. (1975) Phil. Trans. Roy. Soc. London B270, 297-300.
Hodgkin, A.L. and Huxley, A.F. (1952) J. Physiol. 116, 449-472.
Kandel, E.R. (1985) Ch. 11 in "Principles of Neural Science" (2nd edition) E. R. Kandel and J.H. Schwartz (eds), Elsevier, New York.
Kandel, E.R. and Siegelbaum, S. (1985) Ch. 9 in "Principles of Neural Science" (2nd edition), E.R. Kandel and J.H. Schwartz (eds), Elsevier, New York.
Katz, B. (1966) Nerve, Muscle and Synapse (Ch. 7) Mc Graw Hill, New York.
Katz, B. and Miledi, R. (1970) Nature 226, 962-963.
Keynes, R.D. (1986) Ann. N.Y. Acad. Sci. 479, 431-438.
Keynes, R.D. and Rojas, E. (1974) J. Physiol. 239, 393-434.
Koester, J. (1985) Chs. 5-8. In "Handbook of Physiology", vol. I, part I, E.R. Kandel (ed.) American Physiol.

Soc., Bethesda, Md.
Kuffler, S.W., Nicholls, J.G. and Orkland, R.K. (1966). J. Neurophysiol. 29, 768-
Lev, A.A. and Armstrong, W. McD. (1975) Curr. Topics Memb. Tansp. 6, 59-123.
Margineanu, D.G. (1987) Arch. Internat. Physiol. Biochim. 95, 381-422.
Martin, A.R. (1977) pp. 329-355. In "Handbook of Physiology", vol. I, part I, E.R. Kandel (ed.) American Physiol. Soc., Bethesda, Md.
Narahashi, T. (1974) Physiol. Rev. 54, 813-889.
Rayner, M.R. and Starkus, J.G. (1989) Biophys. J. 55, 1-19.
Ritchie, J.M., Rogart, R.B. and Strichartz, G.R. (1976) J. Physiol. 262, 477-494.
Ritchie, J.M. and Rogart, R.B. (1977) Proc. Natl. Acad. Sci. USA 74, 211-215.
Schoffeniels, E. (1980) Advan. Chem. Series, Amer. Chem. Soc. 188, 285-297.
Schoffeniels, E. (1983) Arch. Internat. Physiol. Biochim. 91, 233-242.
Steriade, M. and Llinas, R.R. (1988) Physiol. Rev. 68, 649-742.
Takashima, S. (1976) J. Membr. Biol. 27, 21-39.
Takeuchi, A. (1977) pp. 295-398. In "Handbook of Physiology" vol. I, part I, E.R. Kandel (ed.) American Physiol. Soc., Bethesda, Md.
Thompson, S.H. (1977) J. Physiol. 265, 465-488.
Vasilescu, V. and Margineanu, D.G. (1982) Introduction to Neurobiophysics, Abacus Press, Tunbridge Wells, England.
Zaciu, C. and Margineanu, D.G. (1979) J. theor. Biol. 80, 589-593.
Zucker, R.S. (1989) Ann. Rev. Neurosci. 12, 13-21.

CHAPTER IV
MOLECULAR APPROACHES OF BIOELECTRICITY

Soon after the publication of O. Loewi paper (1921) and the subsequent work done by various physiologists amongst whom Dale stands as a key figure, it becomes more and more apparent that the chemical basis of neurotransmission is certainly the most powerfull hypothesis to explain the synaptic transmission of nerve impulse.

However, some electrophysiologists are not at all convinced and J. Eccles stands, at that time, as the most vigorous and enthousiastic proponent of the electrical transmission. For the proponents of the neurochemical transmission it was also clear, in the Thirties, that two different mechanisms were operating: one for the action potential generation and conduction in the axon, termed a physical process, one at the synapse, a so-called chemical mechanism involving synthesis, and release of the transmitter at the nerve ending, recognition and hydrolysis at the post-synaptic membrane. In contradistinction with such a generally accepted view, was the opinion of David Nachmansohn who, very clearly, perceived the inadequacy of the concepts prevailing at that time. Nachmansohn, a biochemist trained and accustomed to think in terms of enzyme action, was rather surprised to notice a complete lack of biochemical approach to such fundamental question. Thus stimulated by the importance of the problem and the essentially physical interpretation and phenomenological outlook of the electrophysiologists Nachmansohn's interest grew into what would become a life work shaping a completely new field of scientific interest.

Owing to the large amounts of acetylcholine and acetylcholinesterase found in the electric organ of electric fishes, Nachmansohn developped, as early as 1937, the idea that acetylcholine must be the key operative substance in bioelectrogenesis, in keeping with the concept of Meyerhof that the specific operative substance in muscle contraction is ATP.

It is certainly worth remembering that Nachmansohn came across his first electric fish at the World's Fair in Paris in the summer of 1937. The spectacular property of certain species of fishes to produce electricity had been known for a long time already in Roman time. Electric fishes had been used mainly in the XIXe century when sensitive detecting devices available today were not yet invented. The powerfull discharge of these fishes could be measured with less sophisticated instruments. The analogy of the electric organ with a voltaic pile was even recognized, tong in cheeks, by Volta who called his pile an artificial electric organ. And as we know it today, it is even more than an analogy. However little had been done with regard to the biochemistry of electric organ when Nachmansohn started to use it as experimental model after his visit at the World's Fair in Paris. This is when he worked on *Torpedo* in Arcachon just before the onset of the Second World War. As demonstrated by later work electric fishes have proven to be unique tool to further improve our understanding of the molecular aspects of bioelectrogenesis (see for instance Schoffeniels, 1960; Nachmansohn and Neumann, 1975). Very early in his attempts to resolve the contradictions raised by the interpretation of electrophysiological data, mainly those concerning the energetics of the action potential, Nachmansohn has stressed the fact that amongst the cell components, only proteins have the properties that make them particularly well suited to provide proper control for specific cellular properties. Based on thermodynamical analysis performed by Wilson and Cabib on the enzyme acetylcholinesterase, the idea was proposed on firm experimental grounds, that specific ligand recognition could lead to conformational change of the enzyme protein. It was largely exploited in the first modeling of membrane events leading to the action potential. This model has been refined years later when Katchalsky and Neumann in collaboration with Nachmansohn, have introduced newly available informations regarding the properties of the acetylcholine receptor (AchR) and the other components of the acetylcholine cycle. The kinetics of acetylcholine hydrolysis by its specific esterase lead also Wilson to define rather acurately the topography of the active sites and to provide an explanation for the toxicity of organophosphates largely used as insecticides and warfare agents. This lead Wilson to design a potent nucleophilic agent, pyridine-2-aldoxine-methiodide (2-PAM) which would compete with the oxygen of the serine residue, restore the enzyme activity thus repairing the biochemical lesion. 2-PAM has been quickly adopted in many countries as antidote in organophosphate poisoning.

Nachmansohn's approach to the problem of bioelectrogenesis has always resulted in acute conflicts with the views of many electrophysiologists and pharmacologists, the more so that he extended the role of acetylcholine cycle from the synapse, where it is generally accepted, to the generation and conduction of action potential in the axonal membrane, an extension much debated. Thus, it was owing to the seminal influence of David Nachmansohn, that the subject of bioelectrogenesis in axons entered the field of biochemistry. Sure enough, more general aspects of the metabolism of neurons, nerves, axons and even glial cells were studied. The strict dependence of the energy metabolism on a proper supply of glucose and oxygen was soon recognized and the metabolic sequences leading to the synthesis of various transmitter substance were unravelled. This is why it is appropriate to recall first the general characteristics of the metabolism of the nervous system. We shall discuss afterwards the specific biochemistry involved in the generation of the nerve impulse as well as that related to the chemical transmission.

As indicated in the previous chapters, the fundamental importance of lipids in membrane structure is well demonstrated. However, central to this problem is the existence of proteins giving the specific characteristics of the membrane, common to all cells, and some of the unique membrane specializations of nerve cells. Forces acting between lipids and also between lipids and proteins are primarily noncovalent: electrostatic, hydrogen bonding and Van der Waals interactions. Since neuronal membranes are specialized to detect, integrate and transmit signals, we expect to find peculiar biochemical properties to this type of membrane. However basic biochemical properties, common to all cells are found. They are described in standard general textbooks on biochemistry and we shall not go into the details of such a study. Instead we shall concentrate on the biochemical characteristics, as far as they are known today, giving the specific properties to this type of cellular differentiation.

A. Intermediary metabolism in brain

It is often stated that the brain is unique in its high rate of oxidative metabolism. However it must be made clear that it is of the same order as the unstressed heart and renal cortex (Muker and Nicklas, 1978).

The complete oxidation of glucose via glycolysis, Krebs cycle and respiratory chain contribute 36 to 38 mole of ATP (from ADP) per mole of glucose. Approximately 15 % of brain glucose is converted to lactate thus avoiding the Krebs cycle. This can however be matched by a corresponding uptake of ketone bodies. Therefore the net gain of ATP is 33 per mole of glucose utilized. The turnover rate of ATP is high: 50 % in approximately 3 sec. This rate can be higher in certain regions.

The ADP accumulation is prevented by its phosphorylation into ATP and through the operation of the adenylate kinase reaction forming ATP and AMP from ADP:

$$2\ ADP = ATP + AMP$$

AMP is present in small amounts. But a relative small percentage decrease in ATP may lead to a relatively large increase in AMP. Since AMP is a positive effector of the reactions leading to ATP synthesis, such an amplification factor provides a sensitive control for the maintenance of ATP levels (Lowry and Passonneau, 1964).

The level of creatine phosphate in brain is higher than that of ATP and creatine phosphate kinase in extremely active. The creatine phosphate level is controlled by changes in oxygenation thus enabling a control of ATP levels. The ATP formed in the mitochondria is brought to its sites of utilization (membrane, axonal transport, etc...) as creatine phosphate thus the ADP level is sustained and the mitochondrial activity kept under control. The isozyme BB of creatine kinase is characteristic of brain but is also present in other tissues. Glucose is the main carbon and energy source in the brain. Other tissues can of course utilize glucose, but they also rely on the oxidation of fatty acids. This is the case with heart, renal cortex and even liver. During hypoglycemia these tissues stop metabolizing glucose saving more to the brain metabolism. Liver glycogen, the carbohydrate main store, is under neural and hormonal control, thus exhibiting a powerful control favorable to the needs of the brain. That part of the lateral hypothalamus under cholinergic control causes increased liver glycogen synthesis and through the vagus nerve, a decrease in gluconeogenesis.

Stimulation of the ventromedial hypothalamus induces increased glycogenolysis and gluconeogenesis a process mediated by cyclic AMP.

An abrupt disruption in the oxidative metabolism of the brain lead to an immediate arrest of brain function. This is an observation well substantiated that should be related to the fact that the brain extracts, at rest, about 10 % of glucose from the blood i.e. about 28 μmol glucose per 100 g per min. When the blood circulation through the brain slows down, more oxygen and glucose are extracted.

The blood-brain barrier (BBB) restricts the entrance of most water-soluble substances into the brain. And specific mechanisms of transport do exist to carry into brain, important substrates of metabolism. If mannose may keep the brain metabolism at sufficient rate, fructose for instance, cannot because it is not taken up fastly enough. In immature brain, fructose is a proper energy/carbon source due maybe to the fact that immature brain has a lower metabolic need or because the BBB is still incomplete. The glucose concentration in brain is lower than that found in sympathetic ganglia or peripheral nerve terminals (Stewart and Moonsammy, 1966). This is generally explained by the existence of a carrier-mediated mechanism controlling the entrance of glucose into brain. The Km of glucose entrance is 7 to 8 mM in mammalian brain, close to the level of plasma glucose. Therefore, small changes in plasmic glucose may cause significant changes in the amount transported thus, within certain limits, glucose and glycogen concentrations in brain vary with blood glucose concentrations. This in turn explains the acute effects of hypoglycemia. The carrier process involved in nerve terminals has an affinity for glucose some 30 times higher than that of the BBB. Therefore these structures are less affected by hypoglycemia. In vitro, brain cells are sensitive to insulin which increases glucose uptake and glycogen storage. Though insulin has been found in cerebrospinal fluid, and insulin receptors do exist in the brain, there is no evidence of an insulin effect on the intact brain.

Beyond the BBB, brain cells take up glucose rapidly: the affinity of hexokinase for glucose is much greater than the apparent affinity of the capillary transport system. Thus glycolysis is controlled by hexokinase and not by the glucose transport system. However, the influx rate in cerebellum is higher than in the cortex despite the fact that it has a lower glycolytic rate: obviously, influx may be independent of glycolysis in some area of the brain. In other tissues, glycolytic fluxes are controlled by insulin-dependent or insulin-independent processes. In this respect brain is different. Glucose influx becomes important under conditions of increased demand: hypoglycemia below 50 mgr per 100 ml of plasma, seizures, etc.

Brain differs also from other tissues by decreasing rather than increasing glucose transport across the BBB during oxygen lack. This fact could be related to the potential toxicity of lactate. Whatever the exact significance of this observation, it points to the fact that brain functioning is totally dependent of an adequate oxygen supply. It should also be related to the low capacity of mature brain cells for lactate efflux, surely an adaptation protecting the brain from the lactate produced elsewhere during strenuous muscular activity.

In neonate, lactate efflux is much more pronounced, a condition that may explain the relative resistance of immature brain to anoxia.

Glycogen is present in brain in low concentration (3.3 mmol./kg of rat brain). This concentration varies with plasma glucose levels. It is also present in glial cells though restricted to astrocytes in adult brain. In synaptosomes, one finds the enzyme of glycogen metabolism.

Astrocytes glycogen may be a source of energy and carbon for neuron, but direct evidences in favor of this process are still lacking.

There is a rapid turnover rate of glycogen in

brain: about 19 μmol per kg per min. This is approximately 2 percent of the normal glycolytic flux. It is narrowly controlled and suggests that local energy sources are important. However would glycogen be the sole supply, the normal glycolytic flux in brain would be maintained for less than 5 minutes.

In agreement with the general rule of control of the primary metabolism in evolution and differenciation (Schoffeniels, 1971; Schoffeniels, 1984) the kinetic properties of the enzymes of glycogen metabolism are different from those found in other tissues.

The first important characteristics of glycogen metabolism in brain is its relative isolation, due to the BBB, from the metabolic activity of the rest of the organism. The glucocorticoid hormones that penetrate BBB increase glycogen turnover, while circulating protein hormones and biogenic amines are without effect.

However, passed the BBB, local amine levels or drugs that penetrate the BBB and modify local amine levels, membrane receptors etc... cause metabolic changes. UDPG is the substrate of glycogen synthesis. The reaction of transfer of the glycosyl group to the terminal glucose of the nonreducing end of an amylose chain via the formation of an alpha-1,4-glycosidic linkage is the rate-limiting step in glycogen synthesis.

Glycogen synthetase exists in both a phosphorylated (D) form, dependent for activity on G-6-P as a positive effector, and as a dephosphorylated, independent form (I) sensitive to but not dependent on G-6-P. The I form in brain has a relatively low affinity for UDPG. In case of increased energy demand, there is a change from D to I form and an I form with an even lower affinity form for the substrate develops. Thus the inhibition of glycogen synthesis is enhanced, rendering more G-6-P available for glycolysis.

It is interesting to notice that in liver where the glycogen turnover is high, the I form of the synthetase is associated with glycogen formation while in brain the regulation of D form may be responsible for reducing the rate of glycogen formation to approximately 5 % of its potential rate. This interpretation stems out the observation of Goldberg and O'Toole (1969) that the I form in brain is associated with inhibition of glycogen synthesis under conditions of energy demands, while the D form is associated with a relatively small regulated synthesis in resting conditions.

The unphosphorylated form of phosphorylase is in the b form (requiring AMP) for about 10 % under steady-state conditions: it is inactive at the low AMP concentrations prevailing normally. Under conditions of energy demands, there is a rapid conversion to the a form, active at low AMP concentrations.

The kinase of the phosphorylase b is indirectly activated by cAMP and by Ca^{++} released during electrical activity. Ca^{++} is removed from its site of action by endoplasmic reticulum, as in muscle, thus terminating the action of Ca^{++}.

There is no direct evidence that glycogenolysis in vivo is controlled by a switch of phosphorylase to the a form.

Norepinephrine and dopamine activate glycogenolysis via a cAMP production, but epinephrine, vasopressin and angiotensin II act through another mechanism, involving Ca^{++} or a Ca^{++}-induced proteolysis of the phosphorylase kinase.

B. Control of glycolysis

The electrophoretically slow-moving form of hexokinase (Type I) is characteristic of the brain. It exists in a bound form, like in other highly glycolytic cells, to the outer mitochondrial membrane. However, it is also found associated with an as yet unidentified particulate component (Parry and Petersen, 1990). In brain hexokinase has a higher propensy to localize at nonmitochondrial receptor sites.

The energy charge (EC) of the cell, ATP/ADP, controls the activity of brain hexokinase. A decrease in EC induces the binding of the enzyme, therefore increasing its efficiency. The bound form of the enzyme has also a greater affinity for glucose with respect to the competition for ATP generated in the mitochondria. Brain hexokinase is inhibted by G-6-P and ADP, and 3-phosphoglycerate is a negative allosteric effector. ATP and cAMP are also inhibitors. G-6-P, when accumulating, shifts the equilibrium of the enzyme towards a less active state. Therefore these various control mechanisms provide a coherent fine tuning of the activity of the first enzyme of the glycolytic pathway.

The second key regulatory enzyme is phosphofructokinase. It is inhibed by ATP, Mg^{++}, citrate and is activated by $NH4^{+}$, K^{+}, inorganic phosphate, AMP, cAMP, ADP and fructose-1,6-biphosphate.

The glycolytic pathway in brain exhibits a Pasteur effect and the regulatory properties of phosphofructokinase account directly for this. In steady state conditions, ATP and citrate levels are such that phosphofructokinase is relatively inhibited as long as the positive modulators, or disinhibitors, are at low level. The enzyme is quickly activated when the steady state conditions are altered.

At the first step of the triosephosphates,

triosephosphate isomerase favors accumulation of dihydroxy-acetone phosphate, a substrate common to the glycerophosphate dehydrogenase, the action of which is to reoxidize NADH, and lipid pathways.

At rest, when the brain tissue is well oxygenated, it produces nevertheless small amounts of lactate and glycerophosphate thus helping to reoxidize glycolytic NADH. In hypoxic conditions lactate and glycerophosphate increase though the amount of lactate produced greatly exceeds that of glycerophosphate. The five isoenzymes of LDH are present in adult brain with the H form predominant as must be expected for a tissue more dependent on aerobic processes for energy. Under aerobic conditions some lactate is still produced because the reoxidation of NADH is also linked to the various mitochondrial shuttles that may not be fast enough to transfer all the reducing equivalent into the mitochondria.

D-2-phosphoglycerate hydrolase responsible for the production of phosphoenolpyruvate exists as 2 related dimers: one, gamma, specifically associated with neurons, the other, alpha, found in glia. This step requires Mg^{++}, K^+ and Na^+.

The multienzyme complex pyruvate dehydrogenase is controlled in brain, like in other tissues, by its level of phosphorylation. Inactivation occurs by phosphorylation of the decarboxylase through the activity of tightly bound kinase and activated by being dephosphorylate by a loosely bound Mg^{++}-Ca^{++}-dependent phosphatase. About 50 % of the enzyme is active in steady state conditions. Pyruvate and ADP are modulators of the phosphorylation process. Pyruvate and ADP inhibit the kinase. Therefore in a situation where more energy is needed, pyruvate and ADP being increased while ATP and acetylCoA decrease, the complex is more active. Pyruvate dehydrogenase is also inhibited by NADP a situation found in hypoxia. Therefore less acetylCoA is produced thus allowing more lactate to be formed at the expense of glycolytic NADH. NAD may then sustain glycolysis. Pyruvate dehydrogenase activity seems to be the rate limiting step for the entry of pyruvate in the Krebs cycle.

Mitochondria are not permeable to acetyl-CoA. However there is an efflux of citrate in the cytoplasm where it can lead to the production of acetyl-CoA via the activity of ATP citrate lysase.In the presence of choline and choline acetyltransferase acetylcholine synthesis may take place. During hypoxia or hypoglycemia acetylcholine synthesis may be reduced by the decrease of acetyl-CoA production. It should however be mentioned that labelled citrate or acetate are never found in acetylcholine thus leaving the problem of the origin of acetyl-CoA needed for acetylcholine still open.

The acetyl moiety of acetylcholine may also be formed in the synaptosomes, a compartment with high glucose turnover.

The flux of carbon through the Krebs cycle is related to the glycolytic activity and acetyl-CoA production. The activity of the cycle is controlled at various enzymatic steps and by ADP, a well known positive effector of mitochondrial respiration. There are two isocitrate dehydrogenase in brain. The one located in the cytoplasm has NADP as coenzyme. The other, bound to the mitochondria requires NAD and is the enzyme participating in the activity of the Krebs cycle. It is an allosteric enzyme inhibited by NADH and ATP while ADP is a positive effector. The significance of the cytoplasmic enzyme is unclear: it has been suggested that it could provide the reducing equivalents necessary in many reductive biosynthetic processes. The high activity of this enzyme in immature brain is consistent with this interpretation.

The oxidative decarboxylation of 2-ketoglutarate involves a complex closely related to the one responsible for the production of acetyl-CoA from pyruvate and necessitates the same coenzymes.

In brain succinate dehydrogenase has a regulatory role. Notice that succinate levels are little affected by changes in the Krebs cycle activity as long as glucose is available. This is also true for isocitrate.

The free energy change associated with the oxidation of malate into oxaloacetate is unfavorable from the thermodynamic point of view. This situation is overcome by a constant removal of oxaloacetate by condensation with acetyl-CoA. Therefore the concentration of oxaloacetate is always very low.

The malic dehydrogenase is found in both mitochondria and cytoplasm. In the cytoplasm it acts as a member of a so-called mitochondrial shuttle passing the reducing equivalent of cytoplasmic origin to the respiratory chain. It is a complex set of reaction involving:

1) a reduction of oxaloacetate into malate;

2) a transfer of malate in the mitochondria against 2-ketoglutarate of mitochondrial origin via a carrier;

3) the oxidation of malate into oxaloacetate in the mitochondria with production of NADH available for the respiratory chain;

4) the transamination in the mitochondria of glutamate to oxaloacetate leading to the formation of aspartate and 2-ketoglutarate;

5) the exchange, via a carrier of mitochondrial aspartate against cytoplasmic glutamate;

6) the transamination in the cytoplasm of aspartate/2-ketoglutarate forming glutamate and

oxaloacetate. The glutamate thus formed is the one implied in reaction 5) while the oxaloacetate fuels reaction 1).

Therefore the malate-aspartate shuttle requires the cooperation in a cyclic manner of cytoplasmic and mitochondrial isoenzymes of malate dehydrogenase, aspartate-glutamate transaminases as well as the mitochondrial membrane transport system (carriers) for malate/2-ketoglutarate and aspartate/glutamate.

The two other shuttles are the glycerol phosphate shuttle and the acetoacetate/3-hydroxylutyrate shuttle.

The glycerol phosphate shuttle is unidirectional: it transports reducing equivalents into mitochondria.

The complex malate-aspartate shuttle is bidirectional since it can transport reducing equivalents produced by the Krebs cycle from mitochondria to cytosol.

Similarly intramitochondrial reducing power can be used to form extramitochondrial NADPH. Intramitochondrial isocitrate may pass through the mitochondrial membrane on the tricarboxylate carrier to the cytosol, where it may donate electrons to NADP thanks to the cytosol form of NADP-linked isocitrate dehydrogenase.

Therefore the operations of the shuttles and the tricarboxylate carrier for transfer of the reducing equivalents of NADPH effectively regulates the ratio of intramitochondrial and extramitochondrial NADH/NAD and NADPH/NADP couples.

The Krebs cycle functions not only as a catabolic pathway producing through oxidative processes energy, but also as a biosynthetic route providing the keto acids 2-ketoglutarate and oxaloacetate corresponding to glutamate and aspartate (amphibolic pathway). These amino acids are themselves the substrates of other reactions. Therefore it is obvious that a net export of 2-ketoglutarate or oxaloacetate implies a replenishment of the intermediaries of the Krebs cycle via other routes, the anaplerotic pathways.

Certainly the most important route is the one of the carboxylation of pyruvate. Thus the rate of CO_2 fixation sets the upper limit at which biosynthetic reactions can occur.

By studying the acute ammonia toxicity in cats, it has been estimated at 0.15 μmol/g wet weight brain/min or approximately 10 % of the flux through the Krebs cycle. Liver on the other hand, appears to have 10 times the capacity of CO_2 fixation, a fact certainly related to the high amount of proteins synthetised in this organ. In brain pyruvate carboxylase is mainly an astrocytic enzyme.

C. Non oxidative consumption of glucose during neural activity

It is a generally accepted idea that the brain's energy demands under normal conditions is covered by the ATP produced by glucose oxidation (Siesjo, 1978). Thus more than 90 % of all resting-state glucose consumption is oxidative with 5 % or less being metabolized to lactate.

Local cerebral glucose metabolic rate (CMR glu) and cerebral blood flow (CBF) are greatly increased by focal increases in neural activity. The increase in CRM glu has been thought to indicate a local increase in glucose oxidation, supporting large energy expenditures required to maintain the energy and metabolic balances during activity. This would imply a stoechiometric increase in CMRO2 that would be supported by an increase in CBF. There is indeed a large increase (29 %) in local CBF while the increase in CMRO2 is only 5 % (Fox and Raichle, 1986). Therefore if we compare the molar ratio of oxygen to glucose consumption in the resting state or during physiological neural activity, it turns out, as indicated by the results of Fox *et al.* (1988) that it is 4.1:1 in resting conditions and wether we consider whole brain or local areas (visual or somatosensory cortex) while it is 0.4:1 during stimulation.

As expected there is a strong resting-state correlation between CBF and metabolism (CMRO2 and CMR glu) (Table 1).

TABLE 1. CEREBRAL METABOLISM IN RESTING STATE

	CMRO2	CMR Glu	Ratio
mean whole-brain	1.5 ± 0.071	0.37 ± 0.053	4.1:1
primary visual cortex	1.71 ± 0.183	0.42 ± 0.033	4.1:1

Results are in µmol/min/100 g ± SD (after Fox *et al.*, 1988).

During stimulation CMR glu and CBF were markedly increased using a mean of about 50 %. However, $CMRO_2$ increased only by 5 % (Table 2). Thus the molar ratio for the increase in metabolic rate was only 0.4:1

TABLE 2. STIMULUS-INDUCED CHANGES IN THE METABOLISM OF HUMAN VISUAL CORTEX (after Fox *et al.*, 1988)

CBF	CMRO2	OEF	CMR glu
27.1 ± 6.85	0.08 ± 0.19	- 0.17 ± 0.06	0.21 ± 0.037

CBF is expressed in ml/min/100 g of brain tissue.

$CMRO_2$ and CMR glu are expressed in µmol/min/100 g of braintissue.

OEF (oxygen extracted fraction) is the regional O_2 consumption (metabolism to H_2O) expressed as a fraction of the total O_2 delivered by the blood. Values refer to absolute changes, stimulated minus resting state, induced in visual cortex by stimulus.(O_2 glucose). Hence 91 % of the activity-induced increase in glucose uptake was not oxidized.

However tissue pO_2 must have increased during phasic neural activity. Penfield (1937, 1971) noted that local tissue oxygenation, judged by the color of the venous effluent, increased rather than decreased during spontaneous local seizures in human cerebral cortex. Direct measurement of venous O_2 tension during seizures in both animals and humans confirmed this observation (Plum *et al.*, 1968; Brodersen. *et al.*, 1973). Cooper *et al.* (1966) found that a variety of simple motor and sensory tasks produced highly localized increases in cortical O_2 availability. REM sleep also causes large increases in CBF (Reivich *et al.* , 1968; Townsend *et al.*, 1973) and CMR glu (Heiss *et al.*, 1985) but a decrease in OEF (Meyer and Toyoda, 1971; Santiago *et al.*, 1984; Schoffeniels, 1990). In contrast to the minimal increase in O_2 utilization, glucose uptake always rises during phasic neural activity, with a percentage change equivalent to the CBF change (Tables 1 and 2; Fox *et al.*, 1988). This implies that lactate production must increase. This is indeed the case as shown by the results of Hossman and Linn (1987) and Prichard *et al.* (1987). If one assumes that the entire increase in O_2 uptake is used for glucose oxidation at a 4:1:1 ratio and that all of the remaining glucose is metabolized to lactate, the maximum possible increase in ATP production is only 8 % (Fox *et al.*, 1988). One may therefore ask the question to know if the classical view is correct, according to which much energy is consumed by neural activity.

As shown by Fox *et al.* (1988), energy demands during activity is less than 8 %. A similar conclusion is reached by computing from heat production the energy necessary for neural activity. Thus Creutzfeld (1975) estimates that only about 0.3 to 3 % of the cortical energy consumption can be accounted for by spike activity.

A doubling of neural electrical activity should increase CMRO2 by 6 % or less. The enzymatic capacity for glucose oxidation in the brain is near maximal at rest therefore increases of glucose oxidation during stimulation cannot take place (Van den Berg, 1986). All these observations are consistent with the slight increase in $CMRO_2$ during activity as noticed above. Therefore two points remain to be explained. First, how can we account for the increase in CBF?

This increase that accompanies neural activation causes pO_2 and pH to rise and pCO_2 to fall arguing against glucose oxidation as a regulator of CBF under physiological conditions. According to Paulson and Newman (1987) K^+ release during spiking being taken up into astroglial processes surrounding the neuron then released in the blood via the end-feet abutting the capillary could well be the signal regulating regional changes in CBF.

The second question is related to the metabolic fate of this large amount of glucose that is not completely oxidized nor transformed into lactate. One could suggest that the neuron has a much higher capacity for fermentation than is generally assumed. Fermentation processes leading to succinate or propionate, as existing in some Invertebrates have never been investigated. Moreover such processes lead to an extraproduction of ATP (4 extra ATP when fermenting glucose into propionate (Schoffeniels, 1984). This possibility has been discussed in relation to the brain metabolism during the various sleep states (Schoffeniels, 1990). Notice also that the activity of the hexosephosphate pathway so far has not been taken into consideration despite the fact that this metabolic sequence is active in brain as shown in the following section.

D. The pentose shunt

As is well known, the evolutionary significance of the hexosemonophosphate shunt lies in its capability to produce reducing equivalent as NADPH needed for reductive biosynthesis and to provide pentose needed for nucleotide synthesis. Some intermediaries of the cycle are also precursors of some biosynthetic pathway. Thus D-ribose-5-P is a substrate for histidine synthesis, D-erythrose-4-P is a substrate for shikimic acid synthesis, itself a precursor of aromatic amino acids. D-xylulose (or gluconate) may lead to the production of 2-ketoglutarate. Of course the pentose cycle is also an alternate route from hexosephosphate to triosephosphate, paralleling glycolysis. It is misleading to refer to this sequence simply as catabolic or degradative. As is the case with the Krebs cycle, it is an amphibolic sequence, thus indicating their dual roles in catabolism and anabolism. It is then clear that such amphibolic sequence must be regulated by at least 2 inputs. Such a design is logically necessary: a system cannot respond appropriately to two needs unless it receives an input signal relevent to each. It is necessarily true that the amphibolic sequences must respond to at least one signal indicating the level of biosynthetic intermediates and the other indicating the energy level of the cell. When either is low, glycolysis should proceed; when both are in normal range flux through this pathway should be reduced. Dual or multiple control of amphibolic sequences has received little attention as yet and our information in this area is inadequate. Nevertheless it is clear that the controls must be based on the concentrations of small number of compounds that serve as precursors for biosynthesis: sugar phosphates, keto acids, CoA esters and phosphoenolpyruvate. Also since the pentose cycle has the dual properties of regenerating NADPH and of leading to ATP production by providing triosephosphate, it must be regulated by the energy charge of the cell as well as by response to the NADPH/NADP ratio. Therefore in case of ATP needs hexosemonophosphate consumption should increase in both glycolysis and pentose cycle while that part entering the pentose cycle should be regulated by the momentary need of the cell for NADPH. These regulatory interactions are so far tentative and should be taken as a preliminary discussion of correlation of these sequences. Little is known with certainly and the above consideration are simply speculative.

In summary, the direction of flow and the path taken by glucose-6-phosphate after entry into the phosphogluconate pathway reactions are determined largely by the relative requirements of the cell for NADPH and ribose-5-P. If the requirement for NADPH exceeds that for ribose-5-P, the excess pentose phosphate can be converted back into hexose phosphate. If the requirement for ribose-5-P predominates, the flow through the transketolase and transaldolase reactions will be from fructose-6-P to yield pentose phosphates. Only in cells active in reductive biosynthesis (liver, mammary gland, adrenal cortex, brain during myelination) does the complete oxidative pathway leading to NADPH prevail. What are the enzymes that are known to be so far controlled ? The hexose-P isomerase responsible for the entry of glucose-6-P into glycolysis is inhibited by erythrose-4-P. Thus in case of a high demand for NADPH, the intermediates of the pentose phosphate pathway might tend to accumulate. It seems at least plausible that modulation of the isomerase reaction by erythrose-4-P, should prevent such accumulation. Indeed in case of inhibition of the isomerase reaction, the concentration of fructose-6-P would tend to fall. This would drain sugar phosphates from the pentose cycle toward glycolysis and the concentration of all the intermediates should decrease. The glucose-6-P dehydrogenase is also inhibited when the energy charge of the cell or the NADPH concentration are high. Also the kinase of fructose-6-P has a lower activity if the energy charge of the cell is high. These observations make sense from the regulatory point of view since they provide a means to control both NADPH and ATP production. More specifically what is it with the brain ? Turnover in the pentose cycle increases under conditions of increased energy need, for example during and after high rates of stimulation. What could that means? It is important to remember that the C flux through the pentose phosphate in resting conditions amounts roughly to 2.3 % of the brain glucose (Gaitonde *et al.*, 1983). Figures of 5-8 % have been proposed in monkey brain stem, cerebellum and hypothalamus (Hostetler *et al.*, 1970). Oligodendroglia compared with neurons exhibits a relatively higher activity of the hexosemonophosphate shunt enzymes and less of the citric acid cycle enzymes. There is also compelling evidence of the involvement of the hexose monophosphate shunt in synaptic events of neuronal activity (Gaitonde *et al.*, 1983). One should also remember that 2/3 of the C flowing to the pentose cycle enter glycolysis as fructose-6-P, 1/6 of the C enters glycolysis as triose-P and the remaining 1/6 is oxidized to CO_2. In this calculation one does not take into account the use of ribose-5-P and erythrose-4-P as substrates for other synthesis.

One interesting observation is that showing the effects of an inhibition of the pentose cycle. 6-aminonicotinamide inhibits selectively 6-phosphogluconate dehydrogenase. Despite the fact that the contribution of the hexosemonophosphate shunt to the overall utilization of glucose is small, the effects of inhibition of this pathway in the brain are far reaching. The animal treated show several signs of neurological disorder (anorexia, irreversible paralysis, blindness, seizures, etc...). There is also a decrease (16 %) in the overall utilization of glucose, a decrease (26 %) in the labelling of amino acids after injection of (U-^{14}C) glucose and a decrease (10-13 %) in the concentration of Gaba and glutamic acid. There is no clear explanation as to the fact that an inhibition of the hexosemonophosphate shunt accounting for about 2.3 % of glucose utilization could result in a decrease of 16-26 % in the overall utilization of glucose. It shows however an important involvement of this shunt at the level of the brain cells as well as at the synaptic level. Since nearly normal level of NADPH/NADP ratio are found in the brain where the pentose cycle is inhibited, the changes observed are unlikely to be due to a deficiency in NADPH. Further experimental evidence is necessary to throw light on the factors involved and to define the importance of the hexose monophosphate shunt in cerebral metabolism.

One aspect of glucose metabolism in the brain is the tight relations developped with the amino acid metabolism. The activity of the transaminases is such that a large amount of the adequate Krebs cycle intermediates is trapped as amino acids. When labelled glucose is administered, 70 to 80 percent of the radioactivity are found into glutamate and aspartate, 10 to 30 minutes later. This is in contradistinction with the situation found in the liver where the amino acids related to the Krebs cycle are present at much lower steady state values: only 20 % of the radioactivity are found in these amino acids a short time after administration.

This situation is also found in the immature brain and a sharp change occurs during development with the establishment of the metaboic compartmentation of amino acids metabolism characteristic of adult brain.

There are at least two distinct pools for glutamate metabolism in brain. The Krebs cycle intermediates associated with these pools are also distinctly compartmented. The idea now prevailing is that Gaba is metabolized at a site different from its synthesis. The picture one arrives at is that of a small pool of glutamate that flows to a larger pool where it is decarboxylated. It has been shown that glutamate decarboxylase is localized at or near nerve terminals, whereas Gaba transaminase the major degradative enzyme, is mitochondrial.

Evidence also points to the existence of a small glial pool of glutamate. Glutamate released from nerve endings appears to be taken up by glia together with pre- and postsynaptic terminals, converted to glutamine and recycled to glutamate and Gaba. The carbon flow through the Gaba pool is estimated at about 10 % of the glycolytic flux. Part of the Gaba can be recycled to produce succinate, a relevant source of carbon to replenish intermediates of the Krebs cycle.

E. The amino acids pool

The essential amino acids: leucine, isoleucine, lysine, histidine, tryptophane, valine, methionine, phenylalanine, threonine and arginine must derived from diet. Arginine is synthesized by the cell but at a rate unadequate to meet metabolic requirements. It is therefore classified in the category "essential".

The non-essential amino acids are made by transaminations of 2-keto acids: pyruvate (alanine), 2-ketoglutarate (glutamate) and oxaloacetate (aspartate). Glutamate is the precursor of proline; serine is derived from 3-P-glycerate or glycine. Glutamine is synthesized from ammonia and glutamate and asparagine is obtained from transamidation of aspartate.

Ammonia uptake from normal circulatory ammonia levels is sufficient to keep the brain in nitrogen balance.

The composition of the free amino acids pool in the brain is different from that in other tissues: it has a relatively high content of glutamate and related amino acids (glutamine, aspartate and Gaba). Notice that free amino acids are osmotic effectors (Schoffeniels, 1976). Therefore these concentrations are tightly controlled. Notice also that the pool of free amino acids in peripheral nerve is different from that in the brain. Most amino acids in mammalian nerve are lower than in brain. The most important difference is the concentration of aspartate which is higher than that of glutamate. Some compounds such as taurine may be 100 times higher in peripheral nerve than in brain.

The free amino acid pool of the brain undergoes complex changes during the maturation of the brain. The exchange between the free form of amino acids and the ones associated with proteins is rather fast: usually a few hours thus indicating a high turnover rate of the brain proteins. Also one observes high turnover rate for essential amino acids: the half time is of a few minutes. However an increase in the amino acid, concentrations in the blood, for example after a meal, does not alter their brain levels because of the fact that related amino acids compete for the same transport system. A selective increase in one amino acid that is a precursor of a neurotransmitter increases the cerebral level of this amino acid and its metabolic product. This is observed in the case of tryptophan and serotonin (Fernstrom *et al.*, 1974).

F. Concluding remarks

The most salient feature of brain metabolism is the discrepancy between O_2 and glucose uptake. As shown in the preceding pages 90 % of the glucose taken up is not metabolized oxidatively. It is therefore difficult to escape the conclusion that it is used in other metabolic sequence(s). The most obvious candidate would be a fermentation pathway, other than lactate, that could provide the required electron drain as well as an extra supply of ATP. From the consideration of a comparative approach to the energy metabolism in hypoxia, it turns out that succinate and/or propionate fermentation as it exists in some invertebrates could be invoked. Moreover, the fact that the pentose cycle has such an important role in controlling the glycolytic flux, further studies should put more emphasis on the metabolic significance of the C flow through this pathway. It is indeed a shunt of glycolysis since it provides fructose-6-P as well 3-P-glycerate. Therefore depending on the need for pentose or reducing equivalents as NADPH the flux could be diverted to glycolysis or to the hypothetical fermenting pathway we are postulating. Remember that through a carboxylation process the triosephosphate could lead to succinate and propionate. This would add a new dimension to the capacity of the brain to fix CO_2.

When considering the balance of glucose utilization, one has to keep in mind that only 2 % of the glucose flux in whole brain goes toward lipid synthesis and approximately 0.3 % is used for protein synthesis. Therefore the energy demand is mainly related to the turnover of neurotransmitter, to keeping the ionic gradients and to fuel still unknown metabolic sequence(s). Keeping in mind that the maintenance of ionic gradients uses at the most a few percent of the total energy produced and that the energy expenditure related the spike production is also small this leaves a large excess of the glucose uptake unexplained.

References

Brodersen.P,Pauson,O.B,Bolwig, T.G., Regon, Z.E., Rafaelsen, O.J., Lassen, N.A. (1973) Arch. Neurol. 28, 334-338.

Cooper, R., Crow, H.J., Walter, W.G. and Winter, A.L. (1966) Brain Res. 3, 174-191

Creutzfeld, O. (1975) In Brain Work: the coupling of function, metabolism and blood flow in the brain (D.G. Ingvor and N.A. Lassen, Eds.). A. Benzon Symposium VIII, Munksgaard, Copenhagen, pp. 22-47.

Fernstrom, J.D. (1974) Nutritional control of the synthesis of 5-hydroxytryptamic in the brain. In: Aromatic amino acids in the brain (G.E.W. Wolstenholme and D.W. Fitzsimons, Eds.). Ciba Foundation Symposium 22, Elsevier, New York, pp. 153-173.
Fox, P.T. and Raichle, M.E. (1986) Proc. Natl. Acad. Sci. USA 83, 1140-1144.
Fox, P.T., Raichle, M.E., Mintun, M.A. and Dence, C. (1988) Science 241, 462-464.
Gaitonde, M.K., Evison, E. and Evans, G.M. (1983) J. Neurochem. 41, 1253-1260.
Gatfield, P.D. (1966) J. Neurochem. 13, 185-195.
Goldberg, N.D. and O'Toole, A.G. (1969) J. Biol. Chem. 244, 3053-3061.
Hastetler, K.Y., Landau, B.R., White, R.J., Albin, M.S. and Yashou, D. (1970) J. Neurochem. 17, 33-39.
Hosman, K.A. and Linn, F. (1987) J. Cereb. Blood Flow Metab. 7, S297.
Loewi, O. (1921) Arch. Ges. Physiol. Pflƒgers, 189, 239.
Lowry, O.H. and Passonneau, J.V. (1964) J. Biol. Chem. 239, 31-32.
Maker, H.S. and Micklos, W. (1978) Biochemical responses to body organs to hypoxia and ischemia. In: E.D. Robin (Ed.) Extrapulmonary manifestations of Respiratory disease. Dekker, pp. 107-150, New York.
Meyer, J.S. and Toyoda, M. (1971) In: Cerebral circulation and stroke (K.J. Zulch, Ed.) pp. 156-169. Springer Verlag, New York.
Nachmansohn, D. and Neumann, E. (1975) Chemical and molecular Basis of Nerve Activity. Academic Press, New York, 403 p.
Paulson, O.B. and Newman, E.A. (1987) Science 237, 896-
Parry, D.M. and Petersen, P.L. (1990) J. Biol. Chem. 265, 1059-1066.
Penfield, W. (1937) Res. Publ. Assoc. Res. Nerv. Ment. Dis. 18, 605.
Penfield, W. (1971) J. Neurosurg. 35, 124-127.
Plum, F., Posner, J.B., Troy, B. (1968) Arch. Neurol. 18, 1-13.
Prichard, J.W., Petroff, O.A.C., Ogino, T. and Shulman, R. (1987) Ann. N.Y. Acad. Sci. 508, 54-63.
Reivich, M. *et al.* (1968) J. Neurochem. 15, 301-306.
Santiago, T.V., Guerra, E., Neubauer, J.A. and Edelman, N.H. (1984) J. Clin. Invest. 73, 497-506.
Schoffeniels, E. (1960) Arch. Internat. Physiol. Biochim. 68, 1-151.
Schoffeniels, E. (1972) Adaptation at the molecular Scale (Molecular Evolution, E. Schoff. Ed.), Vol. 2, North Holland, 314-335.
Schoffeniels, E. (1976) Biochemical approaches to osmoregulatory processes in Crustacea. In: Perspectives

in experimental Zoology. Vol. 1. Zoology (P. Spencer Davies, Ed.) Pergamon Press, Oxford, pp. 107-124.
Schoffeniels, E. (1984) Biochimie comparÄe. Masson, Paris, 205 p.
Schoffeniels, E. (1990) Arch. Internat. Physiol. Biochim.In press.
Siesjo, B. (1978) Brain energy metabolism. Wiley, New York.
Stewart, M.A. and Moonsammay, G.I. (1966) J. Neurochem. 13, 1433-1439.
Townsend, R.E., Prinz, P.M., Olrist, W.D. (1973) J. Appl. Physiol. 35, 620-625.
Van den Berg, C. (1986) In: Energetics and Human Information processing (G.R.J. Hockey, A.W.K. Gaillard, M.G.H. Coles, Eds.) Nijhoff, Boston, pp. 131-135.

CHAPTER V
PUZZLE OF NERVE IMPULSE THERMODYNAMICS

There is little doubt that making any scientific review is, up to a certain point, a historian's job, but writing now about the energetic aspects of the nerve impulse might appear as a kind of archaeology. However, when dealing with this subject, one is faced with the not so common situation that some undisputed experimental results, namely those concerning the heat evolved by the stimulated nerves, stubbornly resisted the attempts to integrate them in the otherwise successful ionic theory of excitation. The typical human behaviour was to almost forget those results.

In one of his inspired comments about the nature of the nerve impulse, A.V. Hill (1965) wrote: "nerve heat production is rather a nuisance; things would be so much simpler without it". What he meant was that heat production prevents the nerve to function in ideal conditions, but his remark would be equally suited with respect to the theories about nerve impulse. Our convinction is that there are at least two reasons to avoid the temptation of neglecting the energetics of the nerve impulse: 1) no explanation of a natural phenomenon can be accepted until it copes with the laws of thermodynamics and 2) heat dissipation sets practical limitations in high speed computers and knowing how the biological design deals with such aspects might prove to be of use.

V.1. Oxygen consumption and heat production in active nerve

As it was shown in the preceding chapter, nerve metabolism is quite similar to that of other tissues. The common substrate for nerve cells is glucose, its lack causing the progressive decrease of the nerve excitability up to its complete disappearance, in a few hours. Conversely, glucose consumption is increased during and following the electrical activity of the nerve or brain cells.

In nerve, the fermentative metabolic pathways might provide, in anoxia, only about 20 % of the normal energy supply, which is not enough for maintaining the excitability. The loss of excitability caused by anoxia, by lack of glucose or by metabolic inhibitors is due to the decrease of the high energy phosphate supply needed for the operation of the Na^{+}/K^{+} pumps.

In view of the important role played by the experimental model of internally perfused squid giant axon, from which most of the axoplasm is extruded, it is important to remind the demonstration by Martin and Shaw (1970) that in the very thin layer of axoplasm, which still covers the inside of perfused axons, there is ATP synthesis by an enzyme system which is not, as usual, intramitochondrial, but directly located on the internal face of the excitable membrane.

As one could expect, the oxygen consumption of the nerves (which are mere bundles of fibres, specialized for impulse propagation) is much lower than that of the neural centres, which are formed from the neuronal somas, where most of the biosynthetic activity is performed. The data in Table V.1 illutrate these large differences.

Table V.1. Oxygen consumption in $\mu l\ O_2/h.g_{wetweight}$ (data quoted by Vasilescu and Margineanu, 1982)

	Nerve	Centres
Squid	160	950
Frog	270	2400
Cat	≈60	≈450

In the fifties, F. Brink had shown that, when a nerve is stimulated at constant frequency, the rate of its oxygen consumption increases, reaching, in about 30 min, a steady value which depends on the frequency of stimulation. When expressed per impulse, the extra oxygen consumption of the nerve decreases with the frequency of stimulation, from 8 pmole O_2 mol/g at 5 imp/s up to 1.5 pmol O_2/g at 200 imp/s. Later on, Ritchie (1967) confirmed this variation for non-myelinated fibres from rabbit vagus nerve. The resting oxygen consumption of these fibres is 9.24×10^{-8} mol O_2/min.g at 21°C and the stimulation at 3 imp/s causes an extraconsumption per impulse of 8.16×10^{-10} mol O_2/g. This extraconsumption decreases as the frequency of stimulation increases, the dependence indicating that a single impulse produces the consumption of 1.2×10^{-9} mol O_2/g.

As for any other tissue, the measurement of the heat evolved by the nerve is the direct experimental way to set the dissipation of metabolic free energy for all the processes occuring in that tissue. The first successful mirocalorimetric measurements on the nerve were done in 1925 by A.V. Hill. Basically, such measurements imply surrounding the nerve by several hundred thermocouples in series.

In older microcalorimetric investigations, it was only possible to measure the heat production of the resting nerve and the excess caused by a large number of impulses. The difference between the heat evolved by a stimulated nerve, and its heat production at rest is the heat of activity. It consists of two components:

1) the initial heat, which appears and then falls out immediately on and after stimulation, it thus seems directly related with impulse propagation and

2) the recovery heat, which gradually increases upon stimulation and slowly falls down after it ended. The recovery heat forms the largest part (about 97 %) of the heat of activity and it is related to the increased activity of the Na^+/K^+ pumps restoring the composition of the axoplasm. As one could expect, the overall heat production and the oxygen consumption of the stimulated nerves go absolutely in parallel, this proving the metabolic and not simply physical (as mere Joule effect) origin of the heat of activity (Fig. V.1).

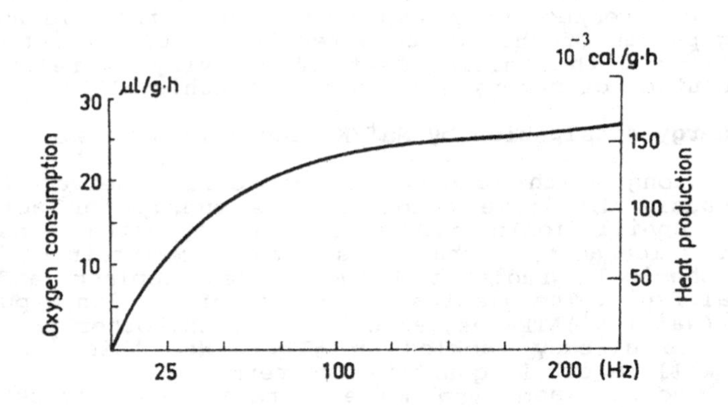

Figure V.1. Oxygen consumption and heat production of a frog sciatic nerve as a function of the frequency of stimulation (from Connely, 1959).

When the electronic equipment became sufficiently fast to follow up the output of the thermopile during the millisecond time domain of the passage of a single action potential, it appeared that the best experimental material are the nonmyelinated nerves, whose fibres have not very different diameters and consequently there is less temporal blurring as a result of nonuniform conduction velocities.

The first modern measurements were on the leg nerves of crab and lobster, then on the rabbit vagus nerve and on the olfactory nerve of pike (data reviewed by Abbot and Howarth, 1973).

The surprising discovery was that, in all cases, the initial heat of activity, represents the difference between an initial heat production - the positive initial heat - and a subsequent reabsorption of most of it - the negative initial heat. Thus, in *Maia squinado* crab nerve, a single impulse gives rise to a production of 8.8 μcal/g, followed by the reabsorption of 6.8 μcal/g, leaving a net initial heat produciton of 2 μcal/g (Fig. V.2). These values vary from one nerve to another, but in all cases about 80 % of the positive initial heat is reabsorbed. Because different chemical agents that affect the time course of the compound action potential of the nerve, modify in a similar fashion the time course of the initial heat, it could be concluded that its positive and negative phases are respectively associated with the rising and falling phases of the action potential, a conclusion which suggests that the initial heat of activity is related to the evolution of membrane potential (Ritchie, 1973).

V.2. Energy dissipation by Na^+/K^+ pumps in nerves

Along with every living cell, nerve fibres continuously dissipate chemical free energy for actively driving uphill ionic fluxes in the opposite sense of passive leakages, that is for maintaining the electrochemical gradients between the axoplasm and the external fluid. The dissipative nature of the ionic pumping by the $(Na^+ + K^+)$ATPases, as well as by any other transport ATPase was already tackled in §II.3. Now this important aspect will be put in quantitative terms.

The axoplasm of a nerve fibre and the interstitial fluid are two subsystems, respectively noted by the superscripts (i) and (e). The overall system is isothermal and isobaric, so that the molar electrochemical potentials are:

$$\mu_k = \mu^\circ_k + RT \cdot \ln a_k + z_k FW \qquad (V.1)$$

The notations are as in §III.1: μ°_k - the standard chemical potential; a_k - the activity of the k ions; W - the electrical potential; z_k - the valency of each kind of ions; R, T and F have their usual meanings. For

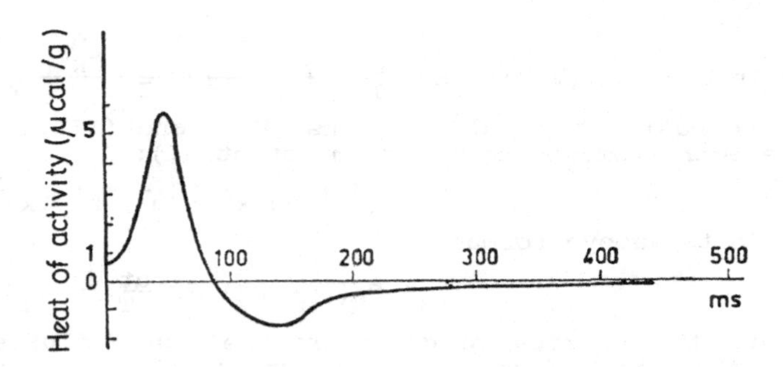

Figure V .2. Heat of activity of a crab nerve, following the conduction of a single impulse (from Hill, 1960).

each of the two subsystems, a similar definition will hold.

Mass conservation for each kind of ion in the overall system implies that the number of moles in the two compartments is constant:

$$d(n_k^i + n_k^e)dt = 0$$

which means the balance of the two opposed fluxes of each ionic species:

$$J_k^i = - J_k^e \qquad (V.2)$$

By definition, the fluxes are taken as dn_k/dt, all quantities refering to unit mass of tissue.

Gibbs equation, incorporating the 1st and the 2nd principles of thermodynamics, gives the entropy variation in the system dS, as a sum of the entropy associated with heat exchanges with the external medium, dQ/T, and that produced by the ionic fluxes within the system:

$$dS = \frac{dQ}{T} + \sum_k \left(\frac{\mu_k^e\, d_k^e + \mu_k^i\, d_k^i}{T} \right)$$

In view of (V.2): $dS = \frac{dQ}{T} + \sum_k \frac{\mu_k^i + \mu_k^e}{T} J_k^i.dt$

and on the basis of the definitions (V.1) and (III.1) (for Nernst electrochemical equilibrium potential):

$$dS = \frac{dQ}{T} + \sum_k Z_k F\, (E_m - E_k)\, J_k^i \frac{dt}{T}$$

In the above formula:

$$\sum_k Z_k F\, (E_m - E_k)\, J_k^i \frac{dt}{T} = d_i S$$

represents the entropy produced by the ionic fluxes. It follows that the dissipation function (Katchalsky and Curran, 1965) associated to these fluxes is:

$$U = T\,\frac{d_i S}{dt} = \sum Z_k F\, (E_m - E_k)\, J_k^i \qquad \text{(V.3)}$$

All the quantities appearing in (V.3) are experimentally accessible, as E_k's are determined by measuring the internal and external activities of the ions with selective electrodes, and the unidirectional flows Jik are measured with tracers. The following values were obtained for power dissipation by the ionic fluxes (Margineanu, 1970; 1977; Schoffeniels, 1989):

$18.3 \times 10^{-6}\ J \cdot s^{-1} g^{-1}$ for frog sciatic nerve

$10.1 \times 10^{-5}\ J \cdot s^{-1} g^{-1}$ for squid giant axon

The physiological significance of these calculations appears from the comparison with the overall energy dissipation in nerves. The resting oxygen consumption of the frog sciatic nerve corresponds to a power dissipation of $2 \times 10^{-4}\ J.s^{-1}\ g^{-1}$ and that of the giant axon, to 9×10^{-4} $J \cdot s^{-1}g^{-1}$. Assuming that the efficiency of the active pumping mechanism is around 50 %, it follows that, in both types of nerves, about 20 % of the energy provided by the oxidative metabolism will suffice for operating the Na^+/K^+ pumps. This conclusion is in full agreement with the fact that ouabain inhibits about 20 % of the resting oxygen consumption in various nerves (Brink, 1975). This shows that the selective block of the active ionic transport causes the diminution of energy metabolism in the same proportion as the estimated relative weight of the transport in the whole assembly of cellular energetics.

During the action potential, the axonal membrane is traversed by large downhill ionic fluxes: an inwardly directed sodium flow and an almost equal outward potassium flow, apart from the much smaller flows of other ions (the "leakage" current). As a result, Na^+ accumulates in the axoplasm and K^+ accumulates outside, so that an increased pumping activity is needed. Upon introducing in (V.3) the values of the ionc flows during an action potential and the time course of potential variation, the energy dissipation due to ionic flows can be computed. The integrated values over the duration of the action potential are (Margineanu, 1972a, 1977):

5×10^{8} J cm^{-2}/impulse for the squid giant axon

6×10^{-7} J cm^{-2}/impulse for the Ranvier node of the bullfrog *X.laevis.A.*

The 12 times larger value for the Ranvier node membrane is related to the ratio of ionic current densities, which are more than 10 times greater in the nodal membrane, as compared with squid giant axon.

A quantity of energy equal to that dissipated by the ionic fluxes which cross the membrane during an impulse, will be stored when the pumps reestablish the ionic gradients. The higher rate of pumping activity is reflected in increased oxygen consumption and heat production (the recovery heat) and as expected, ouabain completely blocks the extra consumption of oxygen associated with nerve activity (Rang and Ritchie, 1968).

It is therefore apparent that, as mentioned in Chap. IV, the energy needed to keep the ionic gradients at rest and during activity, represents a rather small fraction of the total energy available. Thus the use of the remaining energy production has still to be defined.

V.3. Energy changes during the action potential

The above calculations are simple enough for not rising conceptual problems and the agreement with calorimetric and oxygen consumption measurements is actually obvious. However, it is equally true that their only merit is to integrate the ionic pumping in the frame of usual physico-chemical descriptions, with almost no implication for the molecular mechanisms. A rather different situation appears as concerns the thermodynamics of the action potential itself.

It is widely recognized that the nerve impulse is a dissipative, entropy producing process, but the things are neither simple nor obvious in what concerns the source of free energy allowing this dissipation. The many years of experience with internal and external perfusion of squid giant axons have favoured the idea that no solutes other than simple salts and, particularly, no high-energy compounds would be necessary for the propagation of a million action potentials. And, because at some values of the transmembrane potential the ionic channels open and at other values they close in a fully repeatable manner, it was concluded that the electric work done on the gating particles by the external electric field would represent the free energy dissipated during the action potential (Hille, 1978). However, the situation is much more complicated.

The undisputed voltage clamp experiments (§III.3) indicated that, at constant imposed changes in the membrane potential, sodium conductance has a rapid change followed by the spontaneous relaxation to the initial value, in spite of the fact that the transmembrane electric field did not change. Thus, if the state of the potassium channels can be simply field-dependent, it is hard to conceive how this could be the case with sodium channels, which close while the electric field is precisely that one which induced their opening. On the other hand, the energy received by an elementary electric charge, in the form of electric work at a threshold depolarization ($\approx$ 20 mV) is 20 meV = 3.2×10^{-21}J, which is very close to the thermal motion energy $kT \approx 4 \times 10^{-21}$J. Consequently, the stimulus does not appear as a better source of energy than the thermal noise itself, but rather as a promoter of the collective transition in a macroscopic assembly of gateways (Schoffeniels and Margineanu, 1981). Indeed, if one counts the whole electric work done on the nerve membrane by a threshold stimulus, one finds a value much too small to account for the measured thermal changes (see below). Finally, it is to be reminded the previously quoted proof by Martin and Shaw (1970) that the perfused giant axon is not a mere tube devoid of biochemical energetics.

The basic piece of experimental information about the energy changes during the action potential, which has to be accounted for by any mechanistic explanation, is represented by the calorimetric measurements quoted in §V.1. Thus, it is necessary to make clear that what one measures as heat flows (dQ) in isobaric and quasi-isothermal conditions arise from two sources at molecular level. On the one hand are the changes in the intrinsic energy content of the molecular components of the system, i.e. enthalpy changes ($dQ_{p,T}$ = dH). Then, there is the dissipation of the free internal energy that is of the potential forms of molecular energy, because of the intermolecular frictions. In thermodynamic terms, the rate of these entropic heat changes is the dissipation function (V.3). If there exist a whole set of processes, those going downhill are trully dissipative, while the coupled uphill processes diminish the whole dissipation.

The analysis of the origins of the heat of activity of the nerve, particularly of the initial heat of activity, is facilitated if one distinguishes the functional subsystems of the excitable membrane (see also Fig. III.16): i) the lipid matrix of the membrane, to which its dielectric properties are essentially associated, ii) the membrane ionic pumps - the (Na^{+}+ K^{+})ATPases; iii) the passive gateways (conducting sites) through which pass the downhill ionic flows. Apart from, but in structural and functional connection with the membrane are: 1) the ionic compartments (the axoplasm and the external fluid) whose compositions are regulated by the membrane and 2) the chemical pools which fuel the subsystems ii) and, possibly, iii). As for the processes which contribute to the initial heat of activity, they appear to be: a) dissipative ionic interchanges between axoplasm and the interstitial fluid, b) intramembrane redistributions of electric charges and c) enthalpic changes within the conducting sites when passing between open (high conductance state) and closed (low conducting state) conformations.

V.3.a. Ionic dissipation of energy

In the previous section V.2, the overall dissipation of energy by the ionic flows which cross the axonal membrane during the impulse have been given; it represents the energy needed by the Na^{+}/K^{+} pumps to reestablish the electrochemical gradients at their resting values. The time course of the dissipation during an action potential in excess to the resting state, can be computed with the formula:

$$\Phi_{ionic} = \sum_k [g_k(E_m - E_k)^2 - g_{kr}(E_{mr} - E_k)^2] \quad (V.4)$$

in which the subscript k indicates Na^{+}, K^{+}, and leakage currents, the subscript r indicates the resting state and E_m is the instantaneous value of transmembrane potential.

The quantities appearing in (V.4) are given by either Hodgkin and Huxley's (1952) analysis of impulse propagation in giant axons, or by any other mathematical model simulating this phenomenon. The computations given below are based on Dubois and Schoffeniels (1974) data, obtained by numerical simulation of the kinetic equations which describe their molecular model of action potential. The model succeeded in accurately reproducing the variation of the membrane potential during the action potential $E_m(t)$, as well as the variation of the partial conductances $g_k(t)$. The results of computing Φ_{ionic} are given in Fig. V. 3. Because the ionic flows through the axonal membrane during the action potential are downhill, the values of Φ_{ionic} are only positive, thus meaning no storage of free energy within the system, but only its dissipation.

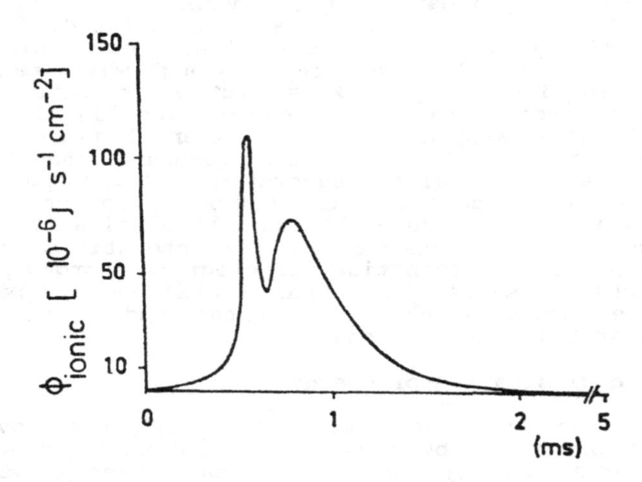

Figure V.3. Energy dissipation by ionic (sodium and potassium) currents through the membrane of squid giant axon during an impulse elicited by a 20 mV depolarization (from Margineanu and Schoffeniels, 1977).

V.3.b. Capacitive energy changes

The dissipation function associated with the variations in charge separation on the two sides of the axolemma during the action potential, Φ_{cap} is the product of the transmembrane potential and the capacitive current: $I_c = C(\partial E_m/\partial t)$, where C is membrane capacity.

It follows that:

$$\Phi_{cap} = -CE_m(\partial E_m \partial t) \qquad (V.5)$$

The minus sign accounts for the fact that a decrease in transmembrane potential corresponds to an evolution of heat, i.e. to a <u>positive</u> heat (in calorimetric terms).

The computed values of Φ_{cap}, are represented in Fig. V.4.

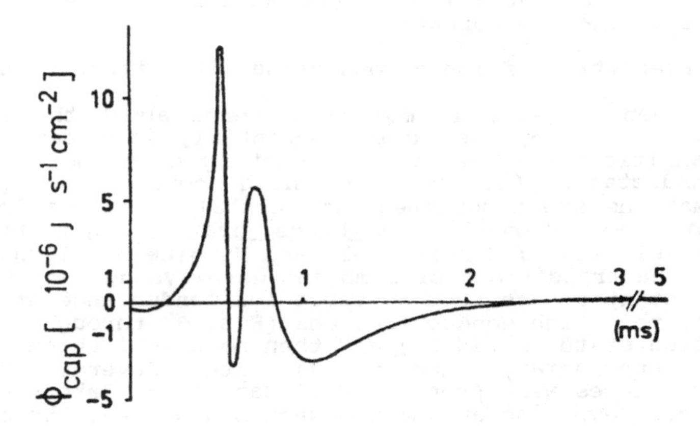

Figure V.4. Capacitive energy changes during the action potential in the squid giant axon (Margineanu and Schoffeniels, 1977).

The overshoot of membrane potential above zero and its return to the resting value of -60 mV are uphill (entropy consuming) processes, which give the negative parts of the graph.

The membrane capacity was taken as constant during the action potential, in accordance with classical indications (Cole, 1970). But owing to the relative smallness of Φ_{cap} compared to Φ_{ionic}, accepting an increase of several tens of percents in membrane capacity (Takashima, 1976), would not significantly change the balance.

After the passage of the impulse, the membrane potential returns to its resting value, the integral of Φ_{cap} over the whole duration of the action potential is zero, but it consists in a heat evolution of 4.12×10^{-2} μcal/g in the first 1 ms, followed by a slower reabsorption of an equal quantity of heat. Because the capacitive heat changes are at least one order of magnitude smaller than ionic heat dissipation, which is always positive, the consideration of only these two types of phenomena cannot explain the biphasic character of the initial heat of activity, as it was experimentally observed. The "condensor explanation" of the initial heat of activity has been generally recognized as incomplete (Howarth, 1975) because it fails to account for the heat absorbed in the falling phase of the action potential.

V.3.c. Energetics of ionic conducting sites transitions

When thinking in molecular terms about the events which occur during the action potential, it appears that the transitions of the ionic conducting sites between the low conductance (closed) and high conductance (open) conformations are associated with enthalpy changes leading to what was termed a <u>membrane heat</u> (Margineanu and Schoffeniels, 1977; Schoffeniels and Margineanu, 1983).

The transitions of membrane gateways during the action potential are from a stable low conductance state (LCS) 1, to a high conductance one (HCS) 3, through a transition state (noted 2) and then back to 1 through one or more intermediary (inactivated) states. Several detailed kinetic schemes were proposed (Goldman, 1975; Jakobsson, 1976), all revolving around the same basic idea. The cyclic nature of the transitions:

$$(LCS)\ 1 \nearrow 2^{\neq} \searrow 3\ (HCS) \swarrow \ldots \nwarrow 1$$

imposes that if the transition LCS → HCS is exergonic, then HCS → LCS necessarily would be endergonic or vice versa.

The spread of excitation along a nerve fibre

implies as a first event the transition of the sodium gateways, so that the rate at which the transition state is attained is proportional to the conduction velocity of the nerve impulse. It then follows that a formal Arrhenius analysis of the temperature dependence of the conduction velocity reveals the activation enthalpy for the passage from the resting to the transition state, from which the sites will pass to the active state. The values of this activation enthalpy for various nerve fibres are listed in Table V.1.

Table V.1. Activation enthalpies (DH]) of nerve impulse propagation calculated from the temperature dependence of the conduction velocities (Margineanu, 1972)

Animal	Nerve	Type of fibre	Diameter (μm)	Conduction velocity at 20°C (m/s)	H (kcal mole)
X laevis	sciatic	myelinated	3	5.4	4.99
X laevis	sciatic	myelinated	22	39.6	4.85
Cat	saphenous	non-myelinated	0.5	0.5	8.55
Squid		giant axon	≈500	50	4.45

Because DH] displays only small variations from one type of nerve fibre to another, while both diameters and conduction velocities are very different, the idea of a unique mechanism of functioning at the molecular level of all excitable membranes is favoured.

The objection against applying the transition state theory to a process so vaguely defined in molecular terms, as is the propagation of action potentials, does not hold when Eyring's theory is used for calculating the standard enthalpies of transition of membrane structural units whose behaviour is reflected in Hodgkin-Huxley parameters. For each kind of membrane subunits (i = n, m and h) undergoing the transition:

$$(1)\ \text{Closed} \underset{\beta_i}{\overset{\alpha_i}{\rightleftharpoons}} \text{open}\ (3)$$

the temperature dependence of α_i and β_i allows calculating the standard enthalpies of transition and this led to values around 10 kcal/mol (Levitan and Palti, 1975). Taking this value simply as a credible order of magnitude, the heat released by one ionic gateway in the transition 1 →·→ 3 and then reabsorbed in the endergonic transition 3

→··→ 1 would be of the order of 10^{-19} cal. The time course of this membrane heat is then given by:

$$\Phi_{mem} = (10^{-19}\ \text{cal})\left(\frac{N_{Na}}{g^0_{Na}} \cdot \frac{\partial g_{Na}}{\partial t} + \frac{N_k}{g^0_k} \cdot \frac{\partial g_k}{\partial t}\right) \qquad \text{V.6)}$$

with: N_{Na} and N_K, the numbers of sodium and potassium gateways per unit area of axonal membrane; g_{Na} and g_K, the partial conductances of the membrane and g°_{Na} and g°_K, their maximal values (§III.3).

Upon introducing in (V.6) the experimental values of NNa, N_K, g°_{Na} and g°_K and the time derivatives of g_{Na} and g_K given by Dubois and Schoffeniels (1974), one obtains Φ_{mem} during an action potential in the giant axon of the squid (Fig. V.5). The integration of Φ_{mem} over the duration of the action potential shows that, before the end of the first millisecond, there is an outburst of 0.026 μJ/cm² or about 0.5 μcal/g, followed by an equal reabsorption of heat during the closing of the channels in the next 2 ms.

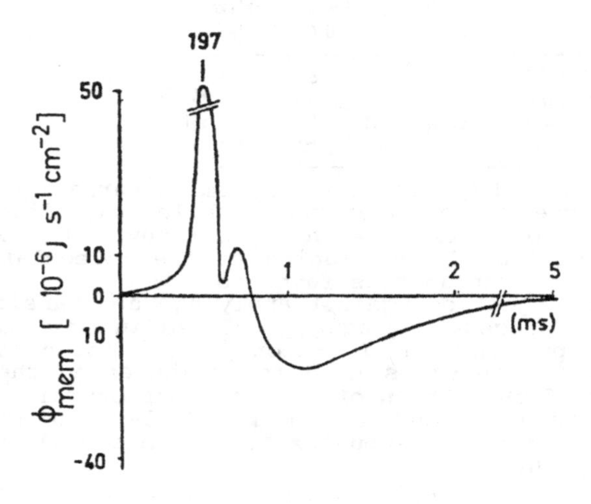

Figure V.5. Energy changes associated with the transition of membrane channels during the action potential in squid axon. The values are computed with eq. (V.6).

The summation of Φ_{ionic}, Φ_{cap} and Φ_{mem}, presented in Fig. V.6 gives an integrated account of the energy changes during the action potential. Obviously, it is still possible that some processes, other than those considered here, might be involved, but it seems likely that the main contributions were included in the calculation. Anyhow, the above developed account of the heat changes does not ascribe the initial heat of activity to only one phenomenon - as it was the case with the unsuccessful condensor explanation - and it proposes the closing of the membrane ionic gateways as the major cause of the negative heat of activity.

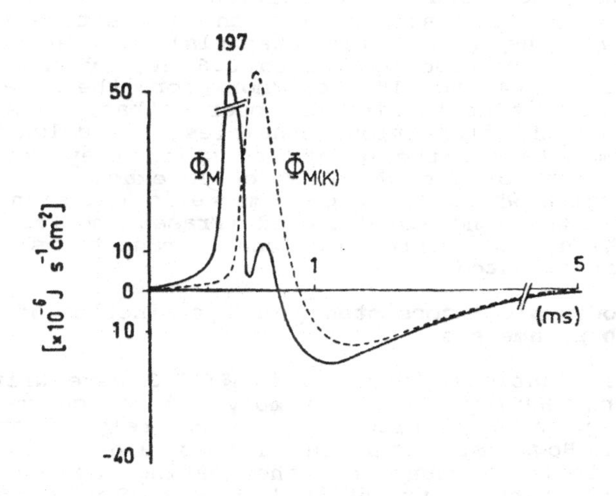

Figure V.6. The sum of computed ionic, capacitive and membrane heats during an action potential in squid axon (Margineanu and Schoffeniels, 1977).

The values in Fig. V.6 show that the rapidity of the thermal changes, together with their small values, prevented obtaining direct calorimetric measurements on the giant axon. Such measurements were successful only in nerves with smaller fibres, such as the leg nerve of the crab *Maia squinado*, the rabbit vagus nerve and the olfactory nerve of the pike, in which the contribution of the capacitive and mostly of the membrane heats rise above the ionic dissipation.

Estimates of heat changes in the case of crab nerves, done on the same lines (Margineanu and Schoffeniels, 1977) reached a fair agreement with the calorimetric measurements and a proposal of pharmacological dissection of the initial heat of activity was put forward. Namely, on the basis of equations (V.4) and (V.6), one could expect that the blockers of sodium channels (e.g. TTX, STX) will eliminate the contributions of sodium to Φ_{ionic} and Φ_{mem}, with major changes in the overall heat changes, but which might not completely disappear. This suggestion is still unechoed.

As concerns the nature of the contributions to the initial heat of nerve activity, one can distinguish between the continuous type of energy, represented by ionic dissipation of heat and the quantified membrane heat, which arises from the transition of some membrane proteins (the ionic channels) between discrete energy states separated by 0.2 to 0.6 eV, if one accepts the values of 4 to 15 kcal/mol for the transition enthalpies. It is comforting to observe that, in spite of the fact that the transition enthalpies were calculated on purely formal bases, the quanta to which they correspond have wavelengths of 2 to 7 μm, which is exactly in the far infrared region where the electromagnetic emission of the active nerve was long ago detected (Fraser and Frey, 1968) and in which the vibro-rotational transitions of the proteins are situated.

V.4. Thermodynamic inconsistency of the kinetics of n, m and h parameters

The equations presented in §III.3 were written by Hodgkin and Huxley (1952) simply as a curve-fitting procedure, with no explicit claim on relying on basic principles. However, later on, Keynes and Rojas (1975) developed their account of the gating currents upon applying Boltzmann distribution theory to the kinetics of gating particles (see also §III.3a). At first sight, this might appear as a wellcome development, but in fact there is an unresolved inconsistency, pointed out by Chapman (1980).

The transition of the subunitary parameter n (defined by eq. (III.19), described by the first order kinetic equation:

$$\frac{dn}{dt} = \alpha_n(1 - n) - \beta_n n$$

indicates that at thermodynamic equilibrium, when dn/dt = 0:

$$\frac{\alpha_n}{\beta_n} = \frac{1}{1 - n} = A_n \exp\,(z_n \theta E_m/kT) \qquad (V.7)$$

In this Boltzman-type of equation, A_n is a constant, k, e and T have their usual meanings (Boltzmann constant, elementary charge and temperature, respectively), E_m is the membrane potential and z_n is the effective valency of each independent quantum of the n system.

Using empirical curve fitting of their voltage clamp data, Hodgkin and Huxley found:

$$\alpha_n = 0.01\ (E_m + 10)/\{\exp[(E_m + 10)/10]-1\}$$

$$\beta_n = 0.125\ \exp(E_m/80) \qquad \text{(V.8)}$$

If one substitutes (V.8) into (V.7), no value of zn can be found satisfying the prerequisite condition of being constant. Similarly, no constant values of z_m or z_n which simultaneously satisfy the empirical Hodgkin-Huxley kinetics and Boltzmann distribution can be found.

Thus, the rate constants appearing in the empirical equations for the conductance systems n, m and h are incompatible with Boltzmann distributions governing the membrane particles supposed to control these systems. The only possibility to reconcile Hodgkin-Huxley description with the basic laws of thermodynamics is to assume that n, m and h particles do not have constant valencies, their charges varying in the course of action potential. By now, neither the mechanism for valency variation of gating particles, nor the energy input for such a process are unambigously known so that we are still in need of a physically realistic model.

x

x x

In brief, the energetics of nerve functioning relies on oxidative metabolism which provides the necessary energy supply for the active transport of ions. This maintains the electrochemical ionic gradients between axoplasm and the external fluid. The free energy stored in these gradients is available to dissipation by downhill ionic currents which propagate the action potential. The amount of energy involved in signal propagation in nerve fibres is really small, of the order of 10^{-6}J/g, per impulse, which makes the energy needs of the nerves to be up ten times smaller than those of the centres. The recorded heat changes associated with nerve impulse propagation can be successfully accounted for if one considers, along with ionic dissipation and capacitive heat changes, the transitions which occur in the ionic conducting sites, when they open and close. This in turn rises the problem of free energy input for the endergonic transitions and suggests that the metabolic supply with ATP of the Na^+/K^+ pumps cannot stand as the only connection between cellular biochemistry and the functioning of nerve fibres.

References

Abbott, B.C. and Howarth, J.V. (1973) Physiol. Rev. 53, 120-158.
Brink, F., Jr. (1975) Proc. Natl. Acad. Sci. USA 72, 3988-3992.
Chapman, J.B. (1980). J. theor. Biol. 85, 487-495.
Cole, K.S. (1970) in "Physical Principles of Biological Membranes", F. Snell et al. (Eds.), Gordon and Breach, New York.
Dubois, D.M. and Schoffeniels, E. (1974) Proc. Natl. Acad. Sci. USA, 71, 2858-2862.
Fraser, A.H. and Frey, A. (1968) Biophys. J. 8, 731-734.
Goldman, L. (1975) Biophys. J. 15, 119-136.
Hill, A.V. (1965) Trails and Trials in Physiology (p. 37), Arnold, London.
Hodgkin, A.L. and Huxley, A.F. (1952). J. Physiol. 117, 500-544.
Howarth, J.V. (1975) Phil. Trans. Roy. Soc. London, B 270, A425-
Jakobsson, E. (1976) Biophys. J. 16, 291-301.
Katchalsky, A. and Curran, P.F. (1965) Nonequilibrium thermodynamics in biophysics (p. 80), Harvard University Press, Cambridge, Mass.
Keynes, R.D. and Rojas, E. (1976) J. Physiol. 255, 154-189.
Levitan, E. and Palti, Y. (1975) Biophys. J. 15, 239-251.
Margineanu, D.G. (1970) Biophysik 6, 327-330.
Margineanu, D.G. (1972a) Kybernetik 11, 73-76.
Margineanu, D.G. (1972b) Experientia 28, 1286-1287.
Margineanu, D.G. (1977) Arch. internat. Physiol. Biochim. 85, 461-478.
Margineanu, D.G. and Schoffeniels, E. (1977) Proc. Natl. Acad. Sci. USA 74, 3810-3813.
Martin, K. and Shaw, T.I. (1970) J. Physiol. 208, 171-185.
Rang, H.P. and Ritchie, J.M. (1968) J. Physiol. 196, 183-221.
Ritchie, J.M. (1967) J. Physiol. 188, 309-329.
Ritchie, J.M. (1973) Prog. Biophys. Mol. Biol. 26, 147-187.
Schoffeniels, E. (1989) Arch. Internat. Physiol. Biochim. 97, 389-402.
Schoffeniels, E. and Margineanu, D.G. (1981) J. theor. Biol. 92, 1-13.
Schoffeniels, E. and Margineanu, D.G. (1983) Topics Bioelectrochem. Bioenerg. 5, 261-305.
Takashima, S. (1976) J. Membr. Biol. 27, 21-39.
Vasilescu, V. and Margineanu, D.G. (1982) Introduction to Neurobiophysics (p. 124), Abacus Press, Tumbridge Wells.

CHAPTER VI
THIAMINE TRIPHOSPHATE AS THE SPECIFIC OPERATIVE SUBSTANCE IN SPIKE-GENERATION.

As shown in the preceding chapters, the action potential is a dissipative process producing entropy and using free energy. This is well demonstrated by:

1) the evolution of the Na conductance under voltage clamping conditions;

2) the microcalorimetric measurements;

3) the analysis of heat evolution during the conductance changes.

The most appropriate explanation must involve an exogenous energy source since the energy dissipated by the ionic flows or even the applied stimulus depolarization are far too small to account for the overall energy balance.

Thiamine triphosphate (ThTP) is a likely candidate as specific operating substance. The more so, since it is specifically hydrolyzed by a triphosphatase the activity of which is modulated by various anions. It is therefore suggested (Schoffeniels, 1989) that the control of Cl^- permeability, a process requiring the hydrolysis of ThTP, is the key to our understanding of the energetics of the action potential.

The interest of neurochemists for thiamine and its phosphorylated derivatives is a long lasting one. It all started with the work of Peters (1936) describing the concept of biochemical lesion after observing the symptomatology of thiamine-deprived pigeon and associating it to the then newly discovery of the properties of thiamine pyrophosphate (ThDP) as cofactor in energy metabolism. However it must be pointed out that despite the general acceptance of Peters suggestion, the precise nature of the involvement of thiamine lack in the production of the characteristic neurological symptoms is still largely unknown. A crucial question to be answered is whether the appearance of these nervous symptoms can be solely explained by the role of this vitamin as coenzyme in intermediary metabolism (decarboxylation of keto acids, transketolase) or whether there exists some more specific properties of this compound in the nervous system (Cooper and Pincus, 1979). This note of caution is even more adequate since it was shown later on that the neurological disorders associated with a deficiency in vitamine B1 were observed well before an alteration of the intermediary metabolism.

The second possibility, which at present is the most widely accepted (for review, see Sable and Gubler, 1982) has arisen, among others, from the following observations.

Electrical activity of nerves is accompanied by release of thiamine-containing compounds (Minz, 1938; von Muralt, 1947). It was also shown that neurotropic compounds such as tetodotoxin (TTX), acetylcholine, ouabain and local anesthetics could also be responsible for release of thiamine either from intact nerves or from membrane preparations (Itokawa and Cooper, 1970a,b,c). These authors showed that this release is preceded by dephosphorylation of ThTP and ThDP, which yields ThMP and Th.

Postmortem analysis of thiamine content in the brain of patients which subacute necrotizing encephalopathy reveals a virtual absence of ThTP (Cooper *et al.*, 1969). These findings together with the fact that ThTP is an integral component of nerve membrane fractions (Itokawa *et al.*, 1972; Schoffeniels, 1983; Bettendorff *et al.*, 1989) as well as of bain synaptosomal membranes (Matsuda and Cooper, 1981) points to ThTP as being the neuroactive form of the vitamin, whereas ThDP, as it is well known, assumes the role of cofactor in the intermediary metabolism of the cell.

Insofar as nerve activity is accompanied by a dephosphorylation of ThTP, it is obviously of interest to study the conditions and properties of the system responsible for the hydrolysis of this compound.

Until now two enzymes have been characterized in rat brain: a membrane-associated (Barchi and Braun, 1972; Barchi, 1976; Barchi and Viale, 1976) and a cytosolic form (Barchi and Braun, 1972) which could be partially purified (Hashitani and Cooper, 1972).

TABLE VI.1. Contents of thiamine and its phosphate esters in different tissues of Electrophorus electricus and the rat.

	ThTP	ThPP	ThMP	Th	Total
E. electricus					
Electric organ	3.9 ± 0.5	0.46±0.08	0.09±0.01	0.06 ±0.01	4.51
Skeletal muscle	0.084 ± 0.009	0.26±0.08	0.32±0.07	0.009±0.002	0.67
Brain	0.37 ± 0.05	4.7 ±0.8	0.12±0.04	0.047±0.002	5.24
Heart	0.047 ± 0.009	2.1 ±0.2	0.35±0.08	0.015±0.002	2.51
Rat					
Brain	0.07 ± 0.01	5.8 ±0.5	0.27±0.07	0.42±0.09	7.16
Heart	0.10 ± 0.03	16.4 ±2.4	1.4 ±0.3	0.17±0.05	18.1
Kidney	0.11 ± 0.01	11.3 ±2.1	2.2 ±0.42	1.3 ±0.1	15.0
Sciatic nerve	0.03 ± 0.01	2.0 ±0.2	0.52±0.20	0.18±0.01	2.73

Data are mean ± SD values, in nmol/g wet weight, from three different animals (after Bettendorff et al., 1987).eel is far higher than those in other tissue. Since the specialization of the electric organ is obvious, it cannot be "a joke of nature" to find so much ThTP in this organ.

Since the electric organ of electric fishes such as Torpedo or Electrophorus are highly specialized in the production of electric current and since the basic phenomena are essentially identical to those found in the brain or peripheral nerves of other species, it is not surprising to find ThTP in these organs. Indeed, the electric organ of Torpedo or of Electrophorus has been found to contain high amounts of ThTP (Eder and Dunant, 1980; Eder *et al.*, 1980; Bettendorff *et al.*, 1987).

In Table VI.1, we compare the values obtained for various tissues of electric eel and rat. It is obious that the amount of ThTP in the electric organ of the

Therefore, it was of interest to look more deeply in the enzymes processing ThTP and to try do decipher some properties of ThTP in this organ or in the mammalian brain.

If one expresses the results of Table VI.1 in percentage, it is seen that the total amount of Th being 100 %, 87 % are in the form of ThTP in the electric organ and 13 % in the skeletal muscle, as far as the eel is concerned. In all the other tissues, from rat or eel, this percentage lies between 0.55 (rat heart) and 7 % (eel brain). It is also worth noting that the ThDP content in the electric organ and the eel skeletal muscle is particularly low when compared to the other tissues listed in Table VI.1.

The electroplax membranes contain a whole set of enzymes responsible for the dephosphorylation of thiamine tri-, di- and monophosphates.

Membrane fractions prepared from this tissue contain a thiamine triphosphatase which is strongly activated by anions and irreversibly inhibited by 4,4′-diisothiocyanostilbene-2,2′-disulfonic acid (DIDS), an anion transport inhibitor. Kinetic parameters of the enzyme are markedly affected by the conditions of enzyme preparation: in crude membranes, the apparent Km is 1.8 mM and the pH optimum is 6.8 but trypsin treatment of these membranes or their purification on a sucrose gradient decrease both the apparent Km (to 0.2 mM) and the pH optimum (to 5.0). Anions such as NO3- (250 mM) have the opposite effect, i.e., even in purified membranes, the pH optimum is now 7.8 and Km = 1.1 mM; at pH 7.8, NO^{3-} increases Vmax 24-fold. ThTP protects against inhibition by DIDS and Kdiss for ThTP could be estimated to be 0.25 mM, a value close to the apparent Km measured in the same purified membrane preparation. ThDP (0.1 mM) did not protect against DIDS inhibition.

At lower (10^{-5}-10^{-6} M) substrate concentrations, Lineweaver-Burk plots of thiamine triphosphatase activity markedly deviate from linearity, the curve being concave downwards. This suggests either anticoperative binding or the existence of bindig sites with different affinities for ThTP. The latter possibility is supported by binding data obtained using (-32P)-ThTP. These data suggest the existence of a high affinity binding site (Kdiss 0.5 μM) for the Mg-ThTP complex.

The membrane-bound thiamine triphosphatase is insensitive to low concentrations of vanadate and is not activated by protonophores. These and other features suggest that this enzyme is quite distinct from most of the known membrane ATPases and that it may have a specific function in excitable cell membranes, possibly in relation to their permeability to certain anions (Bettendorff *et al.*, 1989).

Chloride channels, either ligand-gated (GABA or glycine receptors) or voltage-gated, are inhibited by DIDS. The former does not seem to exist in electric organs (for review see Eldefrawi and Eldefrawi, 1987) and GABA and benzodiazepines are without effect on ThTPase). It is possible that ThTPase and thus ThTP plays a role in the regulation of voltage-gated chloride channels, which are very abundant in Torpedo electric organ (White and Miller, 1979); thiamine compounds are also abundant in Torpedo electric organ (Eder *et al.*, 1976). Chloride channels are important for the stabilization of the membrane potential during the action potential (Eldefrawi and Eldefrawi, 1987; Franciolini and Nonner, 1987). Thus if ThTP controls chloride permeability, it could act as a membrane stabilizer. This view is compatible with the conclusion reached independently by Goldberg and Cooper (1975) and Fox and Duppel (1975), who suggested, without proposing a specific model, a role of thiamine and ThTP as membrane electrical fields stabilizers.

The fact that the activity of thiamine triphosphatase is controlled by anions (Bettendorff *et al.*, 1988) is suggestive of the fact that the free energy needed to the production of an action potential could well be produced by the hydrolysis of ThTP which in turns could control the chloride ions permeability. This possibility has been tested experimentally by measuring in various conditions chloride uptake in microsacs prepared from rat brain (Schoffeniels, 1989).

TABLE VI.2 Effect of GABA and pyrithiamine on the uptake of radioactive choride ion in microsacs prepared from brain according to Harris and Allan (1985).

	Control	GABA (30µM)	Pyrithiamine (1 mM)	Pyrithiamine (0.1 mM)	Pyrithiamine (10 µM)
nmol ^{36}Cl per mg proteins	51 ± 5	79* ± 8	62* ± 9	61* ± 5	55 ± 4

The uptake is measured after 10 seconds.
* probability to the equivalence of the means ($p < 0.05$ (d.d.1 = 10)).

Table VI.2 shows that the uptake of chloride ions is enhanced not only by GABA but also by pyrithiamine, an antimetabolite of thiamine. The results could be interpreted as indicating that the conducting membrane possesses a receptor the occupation of which induces the transconformation responsible for the induction of the chloride flux. Since pyrithiamine has no effect on ThTPase, one should postulate that this ligand keeps its property to activate the receptor while loosing its capability to compete with the substrate (ThTP) of the enzyme. Notice also that ThTP at a concentration of 10^{-3}M activates the chloride uptake after an incubation of 10 minutes (Table VI.3).

TABLE VI.3 Effect of some thiamine derivatives on the chloride uptake by microsacs prepared from rat brain.

Control	Pyrithiamine (100 µM)	Thiamin (1 mM)	Oxythiamin (1 mM)	ThTP (1 mM)
34.5 ± 4.5 (n = 16)	38.8 ± 4.3 (n = 12)	34.6 ± 5.6 (n = 5)	31.7 ± 30 (n = 5)	46.6 ± 9.3 (n = 11)

Choride uptake is expressed in nmol ^{36}Cl per mgr proteins. The compounds are applied 10 minutes before the flux determinations. Figures in brackets are the number of measurements.

Other interpretations of these results are certainly possible but without further experimentation it is difficult to propose a more definite model. It could well be that we are dealing with an exchange Cl^-/HCO_3^- or that the enzyme itself is the conducting site for the anion. Obviously we need to know more about the process before being able to propose a clear picture of the situation. One should indeed remember that the application of neurotropic compounds such as TTX, BTX, STX, Ach, etc...) or ouabain induce a dephosphorylation of ThTP with liberation of Th in the medium. But these compounds have no effect on the enzyme ThTPase. At this stage it is therefore tempting to propose a biochemical cycle of the impedance variation of the membrane comprising the proteins forming the conducting sites for Na, K, Ca and Cl, the Na,K-ATPase, a receptor for ThTP and the ThTPase. Some members of the cycle could be associated through cooperativity relationships since we have to explain the fact that the various compounds mentioned above, if they affect the electrical activity and induce a liberation of Th in the medium are without effect on the enzyme ThTPase.

If the activity of the ThTPase through the control of chloride flux is responsible for stabilizing the membrane potential to its resting value, each time this potential departs from this value, one should observe an hydrolysis of ThTP. Thus the liberation of Th after application of ouabain or BTX could be explained.

But how do we explain the effect of TTX, local anesthetics etc... on the liberation of Th since these compounds do not affect the membrane potential?

The fixation of TTX on the sodium conducting site blocks the transition LCS → HCS. We could well be dealing with an effect of direct cooperativity between the complex protein-TTX and ThTPase: the new configuration of the protein induced by TTX could promote the hydrolysis of ThTP by the enzyme. Or the blocade of the stochastic transition of individual conducting site for Na ions, should modify the dynamics of the membrane in such a way that the enzyme ThTPase would be activated.

Conclusion

Thermodynamical analysis of action potential and more specifically the energetic phenomena associated to the conductance variations taking place in the conducting membrane during an action potential indicate unambiguously that we are dealing with a dissipative free energy consuming and entropy producing process. The conclusion that a substrate must be used to fuel the energetics of the action potential is unescapable and as far as the conducting membranes and particularly the electric organs of electric fishes are rich in ThTP, which is dephosphorylated during electrical activity together with a thamine production in the medium, it seems reasonable to consider ThTP as the specific operative substance in bioelectrogenesis.

The fact that a ThTPase does exist in the

conducting cell associated to the fact that this enzyme is controlled by anions bring support to the idea that the hydrolysis of ThTP provides the necessary energy for the production of an action potential via the control of the chloride permeability.

References

Barchi, R.L. (1976) J. Neurochem. 26, 715-720.
Barchi, R.L. and Braun, P.E. (1972) J. Biol. Chem. 23, 7668-7673.
Barchi, R.L. and Viale, R.O. (1976) J. Biol. Chem. 251, 193-197.
Bettendorff, L., Grandfils, Ch., Wins, P. and Schoffeniels, E. (1989) J. Neurochem. 53, 738-746
Bettendorff, L., Michel-Cahay, C., Grandfils, Ch., De Rycker, C. and Schoffeniels, E. (1987) J. Neurochem. 49, 495-502.
Bettendorff, L., Wins, P. and Schoffeniels, E. (1988) Biochem. Biophys. Res. Comm. 154, 942-947.
Cooper, J.R. and Pincus (1979) Neurochem. Res. 4, 223-239.
Cooper, J.R., Itokawa, Y. and Pincus, J.H. (1969) Science 164, 74-75.
Eder, L., Hirt, L. and Dunant, Y. (1976) Nature 264, 186-188.
Eldefrawi, A.T. and Eldefrawi, M.E. (1987) FASEB J. 1, 262-271.
Fox, J.M. and Duppel, W. (1975) Brain Res. 89, 287-302.
Franciolini, F. and Nonner, W. (1987) J. Gen. Physiol. 90, 453-478.
Goldberg, D.J. and Cooper, J.R. (1975) J. Neurobiol. 76, 2620-2624.
Harris, R.A. and Allan, A.M. (1985) Science 228, 1108-1110.
Hashitani, Y. and Cooper, J.R. (1972) J. Biol. Chem. 247, 2117-2119.
Itokawa, Y. and Cooper, J.R. (1970a) Biochem. Pharmacol. 19, 985-992.
Itokawa, Y. and Cooper, J.R. (1970b) Biochim. Biophys. Acta 196, 274-284.
Itokawa, Y. and Cooper, J.R. (1970c) Science 166, 759-760.
Itokawa, Y., Schulz, R.A. and Cooper, J.R. (1972) Biochim. Biophys. Acta 266, 293-299.
Matsuda, T. and Cooper, J.R. (1981) Proc. Natl. Acad. Sci. USA 78, 5886-5889.
Minz, B. (1938) C. R. Soc. Biol. (Paris) 127, 1251-1253.
Muralt, A.V. (1947) In Vitamins and Hormones, Vol. 5 (Harris, R.S. and Thimann, K.V., Eds.), pp. 93-118. Academic Press, New York.
Peters, R.A. (1936) Lancet 1, 1161-1164.
Sable, H.Z. and Gubler, C.J. (1982) Ann. N.Y. Acad. Sci. 378, 1-468.

Schoffeniels, E.(1989) Arch. Internat. Physiol. Biochim. 97, 389-402.
White, M.M. and Miller, C. (1979) J. Biol. Chem. 254, 10161-10166.

CHAPTER VII.
MERGING ELECTROPHYSIOLOGY AND MOLECULAR APPROACHES

The electrophysiological studies, although essentially at a macroscopic level, were able to settle the problem of how the specialized membrane proteins, through which the alkaline ions traverse the axonal membrane, operate and are able to discriminate between very similar ions, such as Na^+ and K^+.

Different authors, chiefly, Hille (1984) have drawn functional "robot-portraits" of the ionic channels. These are viewed (Fig. VII.1) as transmembrane protein aggregates, in the lipid bilayer and embedded anchored to other membrane proteins, or to elements of the cytoskeleton, which severely restrict lateral movements and confine them to very limited membrane domains, such as the Ranvier nodes. Channel proteins form water-filled pores, much wider than an ion over most of their length and having atomic dimensions only in a short stretch, <u>the selectivity filter</u>, where the ionic selectivity is established. The pore wall is lined by hydrophilic amino acids, while the hydrophobic amino acids interface with the lipids. Gating would result from conformational changes that move a gate into and out of an occluding position. The probabilities of opening and closing are controlled by a <u>sensor</u> which, in the case of voltage-operated channels, consist in charged groups that move in the membrane electric field.

A detailed investigation of the permeability of the sodium channel for cations with different Van der Waals radii, led Hille to conclude that sodium passes complexed with one water molecule and that the selectivity filter possesses groups bearing an ionized oxygen with which the complex Na+.H2O can establish hydrogen bonds. The selectivity filter of sodium channels is a 3.1 Å/5.1 Å rectangle lined by 6 oxygen atoms. In view of its diameter, potassium could pass through sodium channels when it is not hydrated, but in this case it cannot form H-bonds with the oxygens, while when hydrated, it is too large to enter the structure.

For the potassium channels, Hille proposed cylindrical pores with diameters 3-3.3 Å. Na^+ and Li^+ cannot pass through these channels because the strength of the electric field is too small, so that these ions do not loose their hydration water molecules (more strongly attached than with K^+) and thus they are too large.

The functional characteristics of the ionic channels, derived from electrophysiological, macroscopic experiments, can be now substantiated by direct measurements, still electrophysiological, but in fact at molecular level, which were made possible by the patch-clamp

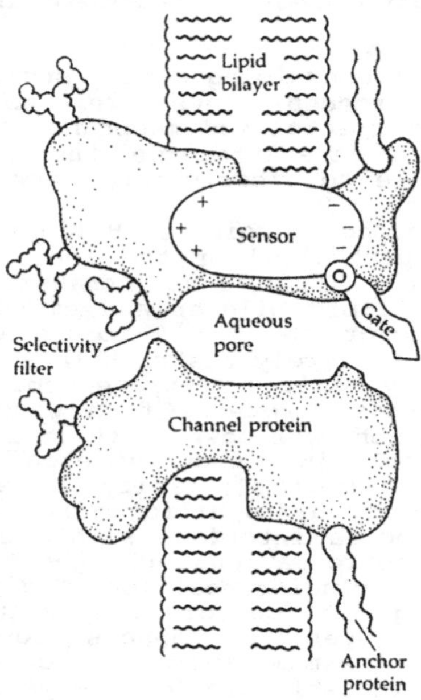

Figure VII.1. Hille's (1984) functional image of the gated ionic channels, such as the sodium channels in the axonal membrane.

technique. Allowing current measurements on single channels, the patch-clamp is the first method able to investigate the function of one protein, left in its natural milieu.

VII.1. Single-channel recording

In order to distinguish the contribution arising from a single channel, the membrane area under investigation must be restricted to less than 30 μm^2. Neher and Sackmann (1976) were the first to achieve this, by gently pressing the tip of a glass microelectrode onto the surface of a muscle cell membrane. Thus, a small portion of the membrane is electrically isolated, the resistance of the seal established between the glass electrode and the membrane being in the range of 10^{-50} MΩ. For reducing the recorded noise, the seal resistance has to be of the order of, but greater than the resistance of the channel itself, which is around 10 GΩ. By heat polishing the electrode tip and by applying a 20 to 30 cm H_2O suction to the pipette interior after the contact with the cell surface, seals of very high resistance (1-100 GΩ) can be obtained (Hamill *et al.*, 1981). Such a tight seal results from ionic bonds between positive charges on the membrane and negative charges on the electrode surface, hydrogen bonds between nitrogen atoms of the phospholipids and oxygen atoms on the glass surface, salt bridges between negative charges from both surfaces, mediated by divalent cations (Ca^{++}) and Van der Waals interactions.

The high resistance of the seal ensures recording of only the currents which pass through the channels in the patch. Besides this, the membrane patch can be easily voltage clamped by fixing the potential of the pipette interior versus that of the external bath, as the intracellular potential remains constant.

The measured channel properties are the conductance and the selectivity towards various ion species, as well as channel kinetics between its closed and open states. The first two parameters are the most basic characteristics of a channel, as they reflect <u>which ions</u> are transported (the selectivity) and <u>how fast</u> this is done (the conductance).

The conductance and selectivity of the channel are determined from a plot of current voltage relationship. From its slope, the channel conductance is directly estimated and, from this, the effects of various exogeneous agents can be assessed. As for the selectivity, it can be characterized by measuring the reversal potential, - the transmembrane voltage at which there is no net flux of ions through the channel. If the ionic gradients across the patch are known, the values of the reversal potential can be used to estimate the channel selectivity. In the case of a channel perfectly selective for one ion species, the reversal potential will be given by the Nernst electrochemical equilibrium condition for that ion.

The channel kinetics is inferred on the basis of sufficiently long recording of opening and closing of the same channel. From such record, time histograms are constructed, where the relative (i.e. percent of total) number of openings that last a time interval t is plotted against t. These histograms are then fitted with linear combinations of exponential functions. For one open and one closed state, the corresponding frequency histogram will be fitted by a simple exponential (Colguhoun and Hawkes, 1983).

$$F(t) = \exp(-\lambda t)$$

where λ is the rate constant of the transition between the two states. This constant results from the fit of the open-time histogram as an experimental value, which indicates the mean length of time the channel stay open: $\tau = 1/\lambda$.

The exponential probability function of the time durations of the open state results (for instance) from the following reasoning in molecular terms. An open channel is a protein in a certain conformation, its bonds vibrating on the picosecond time scale. Each stretch of the molecule is trial to overcome the energy barrier which separates the open state of the channel from the closed one. The probability of shutting at each trial is small, so that the Poisson distribution of the rare events holds. This means that the probability of the channel shutting x times in the interval t is:

$$P(x,t) = \frac{1}{x!} \left(\frac{t}{\tau}\right)^{x} \cdot \exp(-t/\tau)$$

where t/τ gives the mean number of shuttings (because t/τ is taken as the mean lifetime of the open state). Accordingly, the probability of having no shutting, i.e. of remaining open will be:

Before closing this brief quotation of the patch-clamp method, we insist on the fact that it gives in a direct way such individual channel characteristics as the conductance and open time.

VII. 2. The structure of channel proteins

In the Eighties the technical progress in protein chemistry and in pharmacology allowed the extraction, purification and characterization of many channel proteins starting with the postsynaptic ACh receptor, but also involving the voltage gated sodium channel. It was purified from the electric organ of Electrophorus fishes, the rat brain (Elmer *et al.*, 1985), or from skeletal muscle sarcolemma (Barchi, 1983). All the Na channel preparations, contain polypeptide of MW around 260 kDa, which has nearly 500 sugar residues as oligosaccharide chains covalently attached to about two thousand amino acid residues.

These preparations have the major characteristics of Na channel, as they bind TTX and scorpion venoms and, when reincorporated into phospholipid vesicles, they induce ionic fluxes with selectivity for sodium and are stimulated by veratridine and batrachotoxin.

Within the last years, the recombinant DNA technology provided an essentially new approach to study the primary structures of the ionic channels (Lester, 1988). The primary structure of the sodium channel from the electric organ of *Electrophorus* was established by cloning and sequencing the complementary DNA (Noda *et al.*, 1984), then the same approach led to deducing the primary structures of two types of sodium channels from rat brain (Noda *et al.* 1986).

These proteins consist of 1820, 2009 and 2005 amino-acid residues, respectively. They present four internal repeats with homologous sequences referred to as repeats I, II, III and IV. Each internal repeat has five hydrophobic segments (S1, S2, S3, S5 and S6) and one positively charged segment. The hydrophobic segments S1-S6 in each repeat presumably traverse the membrane, forming α-helical structures. The segment S4 contains four to eight arginine or lysine residues, positively charged at their free amino groups. Because of having (+) charge, this segment is unlikely to constitute the inner wall of the channel, which must be lined by (-) charges in order to be selective for Na^+. The structure of segment S4 is nearly identical in the three sodium channels and it appears quite plausible that its positive charges represent the voltage sensor of the channel. With the knowledge of the primary structure, the intramembrane topology of sodium channel appears to be as in Fig. VII.2.

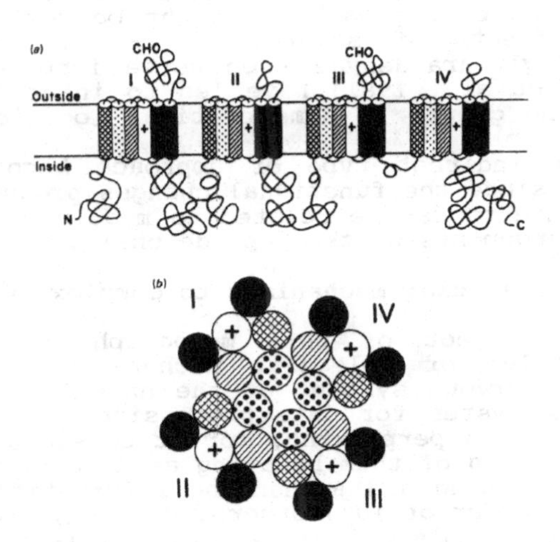

Figure VII.2. Transmembrane topology of the voltage gated sodium channel (a) and the arrangement of the transmembrane segments viewed perpendicularly to the membrane (b)

according to Noda et al., (1986).

Recently, a small protein (130 amino acids) was cloned that induced a slow voltage-gated potassium current (Takumi *et al.*, 1988). It might be involved in the permeability of nonexcitable epithelial cells to K^+ ions, but differs from potassium channels in the nerve cells. The transmembrane protein complex that contains bot the ACh binding site and the cationic channel in the postsynaptic membranes was the first whose primary structure was elucidated (for review see Numa, 1986). Even more, its structure (at 17 Å resolution), was recently determined by three-dimensional image reconstruction from tubular vesicles having ACh receptors organized in helical arrays (Tokoshima and Unwin, 1988)

The knowledge of the amino acid sequence of Na channels revived the interest toward the functional consequences of *in situ* chemical modification of excitable membranes by means of group-specific reagents. Thus, the modification of sodium and of the gating currents by amino group specific cross-linking confirmed that lysine residues are critically involved in sodium channel gating (Drews and Rack, 1988), as suggested by the above mentioned structure of S4 segment and its supposed gating role. In connection with this kind of approach, it might be reminded that the clear-cut effects of several cross-linking aldehydes - chiefly the glutaraldehyde - on nerve impulse propagation (Margineanu *et al.*, 1981) have led to infer the existence of free amino groups in some critical locations of sodium channels.

That indirect type of approach becomes now more attractive, since the functional changes produced by group-specific reagents can be related to models of the possible spatial arrangements of the peptide chain.

VII.3. From molecular mechanisms to complex brain functions

The subject of this monograph is part of the general problem of molecular mechanisms underlying the function of nervous systems of the animals, including the most complex system for data processing, the human brain. Its complexity is perplexing, for it consists in some 10^{12} nerve cells, each of them behaving as a complete system for receiving, treating and exchanging information from and to many (of the order of 10^4) others. However, brain functions rely on a few simple principles whose understanding allows the analytical approach of the cognitive and integrative functions.

The key tenet of contemporary neurobiology is that mind represents the ensemble of functions carried out by the brain, the different types of behaviour emerging from the interconnections of nerve cells. Both the cells themselves, and their connections are subject to modification by experience, but they always behave as signaling units with essentially similar mechanisms which are the same in all animals. Independently of size, shape, transmitter biochemistry and specific behavioral function, every neuron can be described by a model with four components (Fig. VII.3).

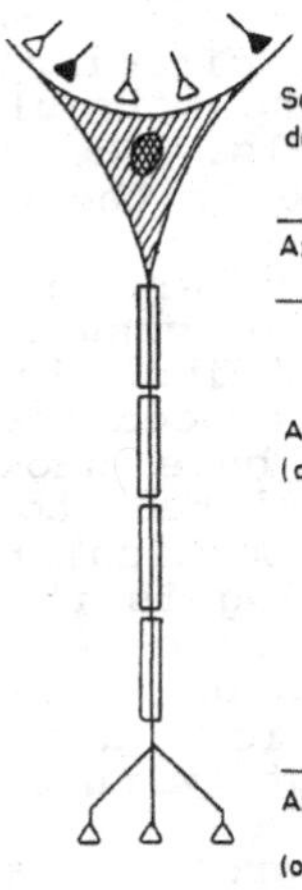

Zone	Function	Potentials
Somato-dendritic	Integration of excitatory (△) and inhibitory (▲) inputs	Graded postsynaptic and receptor potentials
Axonal hillock	Triggering	All-or-none action potentials
Axone (conducting zone)	Propagation	
Axonal endings (output zone)	Transmitter release	Secretory potentials

Figure VII.3. The four functional parts of a neuron and their distinctive features.

The <u>receptive part</u> is that one where the input from other nerve cells or from environmental stimuli is received. In central neurons, such as motor neurons, the input consists in the synaptic potentials elicited by the chemical transmitter, released by other neurons. In sensory neurons, the input is represented by the receptor (or generator) potential which transforms the sensory stimuli of different types (mechanical, thermal, etc.) into a change of the resting potential of the nerve cell membrane. Both synaptic and receptor potentials are graded, of small amplitude (between 0.1 and 10 mV) and propagate passively by the electrotonic mechanisms described in §III.2b. Thus, they cannot be faithfully transmitted, but they summate, initiating in the trigger zone of the neuron the conducting signals - the action potentials which propagate regeneratively (see §III.3) as all-or-none signals along the axon. The trigger zone is the initial segment of the axon where the threshold for action potential initiation is minimal. At the axonal endings of the neuron, the action potential acts as stimulus for the secretion of chemical transmitters. This release is the output of most neurons, except for those which have electrotonic synapses.

In spite of some variability, within this general scheme, the electrical signaling properties of all neurons are stereotyped, which makes for instance that the action potentials coming into the nervous system through sensory axons are indistinguishable from those carrying the motor commands from the brain. The specificity of signals is given by the pathways they are circulating, not by peculiarities of action potentials. The universality of this bioelectrochemical event made it the central theme of this monograph.

The different signaling properties of the various parts of neurons arise from the nonuniformity in the distribution of ionophoric proteins in the neuronal membrane. The immunocyto-chemical procedures and the use of fluorescent labeled tetrodotoxin have shown that voltage gated sodium channels are restricted to some distinct areas, the Ranvier nodes being particularly "hot spots" (see §III.2). The nonuniform mapping is the rule for other channels too, which in particular accounts for the special organization of the ionic channels of some central neurons endowing them with either electrical auto-rhythmicity or with resonance preference for inputs with given frequencies (Llinas, 1988).

Because membrane channel proteins are continuously replaced without alteration of their basic pattern of distribution, it must exist a strict cytoplasmic control that might involve the cell nucleus. Maintaining given types and quantities of channels at specified sites involves differentiated gene expression and replacement of channels, carried to the membrane via the cytoskeletal system. The two-way communication between the plasmalemma and the nucleus is a reasonable basis for understanding long term memory and modulation of behaviour.

The terms learning and memory have been used in such different senses, that it is necessary to define them in each case. Learning is the tendency an organism has acquired as a result of training to respond with increased probability in a given manner, when several alternative actions are possible. As a rule, memory means only the processes of storage and retrieval or "reading" of information. These definitions are not specific to human learning and memory, but apply to the behaviour of any animal. Indeed, in contrast with the older anthropocentric conception, learning phenomena have been found at any level of the animal kingdom where they have been looked for (see instance Mpitsos et al., 1978). In man, learning processes and memory have a qualitatively higher development, even when compared with the nearest primates, as a result of the existence of the second signalling system. Perhaps one of the factors generating the idea of "soul", as something lasting longer than the body, is the insensitivity of memory to treatments strongly affecting the functions of the organism, since memory, once established, is practically impossible to abolish.

There are two very different phases of memory: *short-term memory* (STM) - lasting for only several minutes - and *long-term memory* (LTM) which can last for the whole life. STM consists in a specific type of electric activity induced in the brain during learning, as for instance the facilitated circulation of impulses in reverberating neuronal networks. The essential argument supporting the electrical nature of STM is that the treatments which interfere with electrical activity - electroshocks and chemical agents which produced generalized convulsions, anaesthesia, hypoxia and hypothermia - all abolish STM, but none of them has any major effect on LTM.

As far back as 1911 Ramon y Cajal associated *learning and memory* with synaptic activation and development, and *amnesia*, with a regression in synaptic efficiency following reduced utilization. In 1951 Young put forward the idea that a mature synapse is the culmination of continuous processes of growth and degeneration of neuron terminals, and he suggested the idea of modifications of these processes by neuronal electric activity as the basis of nervous system plasticity. More recently, Changeux and Dauchin (1976) assumed that the formation of neuronal networks is based on the selective stabilization of synapses during development. They postulated the existence of genetic determinants which regulate the connexions between classes of cells but suggested that the final interconnexion pattern depends on the selective stabilization of certain synapses during neuronal activity. The mechanisms of these processes are now less hypothetical, due to the study of elementary types of learning in simple animals such as the marine snails *Aplysia* and *Hermissenda* (Alkon, 1988).

Such a simple type of associative learning is the Pavlovian conditioning, in which an organism learns to link two discrete stimulus elements (just as Pavlov's learned to associate the smell of meat with the ringing of a bell). Thus for instance, *Hermissenda* - which in nature responds to an increased turbulence of ocean by flexing its muscular foot in order to anchor, itself to a support - learns to do so in response to a flash of light, after this conditioned stimulus was associated one to hundred trials with the unconditioned one (water turbulence). Alkon (1987) provided evidence that the mechanisms of associative memory have been conserved over the course of evolution, as this is quite similar in very different species.

Repeated temporal association of stimuli cause a persistent reduction of the flow of potassium ions through the membranes of the target neurons. As it was shown in Ch. III, potassium-ion flow keeps the membrane potential below the threshold so that a reduction of this flow makes the impulses to be triggered more readily. The reduction of K^+ flow lasts for days, which is a time domain suited for storing the temporal relation of the stimuli. This appears to result from the movement of two calcium-sensitive enzymes protein kinases from the cytoplasm to the membrane, where they reduce K^+ flow. Then it was found that efficiency of memory storage is closely correlated with increases in the synthesis of several species of mRNA, the molecular precursor of protein. Five days after the training of snails, the branches of the neurons involved in the processing of conditioned stimuli have a much more condensed volume than in control animals. The conclusion is that a pattern of stimulation is represented and stored by a physical pattern of branching and of synaptic contacts, after it was successively represented and stored by a pattern of electrical signals and a pattern of molecular activation of the protein kinases. While the input information from the unconditioned stimulus flows within the neural system through pathways that are genetically determined or hard-wired, that one from the conditioned stimulus flows along a new pathway, one that is not hard-wired, but formed in the course of the learning experience.

Learning and memory are functions which occur on a macroscopic time scale (with the day as time unit) as they belong to the macroscopic neural system. Their much longer duration as compared with simple membrane phenomena (see §II.2) is due to the involvement of the much slower protein synthesis processes.

VII.4. Levels of unitary events

The ultimate reason for speaking about events, and also for the appearance of fluctuations, is the discrete and quantal nature of matter, energy, and electric charge. With more specific reference to membrane transport, discrete events are represented by the passage of single molecules through the membrane, but the existing experimental procedures almost exclusively allow detecting the manifestation of the particulate nature of the electric charge. Indeed, no physico-chemical analytical procedure has approached so far the limit of detecting the presence of only one molecule, while electrical measurements on membranes are able to reveal several levels of discrete events, connected with membrane transport.

The macroscopic electrophysiological records from multicellular neural centers, such as the resting electroencephalograms, appear noisy due to the stochastic firing of individual neurons; this might be considered a cellular level noise (Fig. VII.4a).

The miniature post-synaptic potentials, such as the miniature end-plate potentials (mEPPs) are local potential changes of less than 1 mV amplitude, which randomly appear in the absence of stimulation with mean frequencies around 1 s^{-1}, due to the spontaneous release of the quanta of chemical mediator contained in the presynaptic vesicles. The random and quantal character of mEPPs, governed by the Poisson law of rare events, is imposed by the quantal release of chemical mediator contained in the presynaptic endings (see §III.4a). Whether or not the quantal release of chemical mediators is a general feature of synaptic transmission is out of the scope of this discussion, what actually matters being that, at least in some cases, a subcellular level noise appears (Fig. VII.4b) as a result of the release at once of some 104 transmitter molecules.

When Katz and Miledi (1970) discovered that the depolarization produced by a steady dose of chemical mediator applied to an end-plate is accompanied by a significant increase in voltage noise (§III.4b, they revealed the membrane level noise P (Fig. VII.4c), produced by the random openings and closings of those channels in the post-synaptic membrane, which are opened by the transmitter. The elementary event is in this case, as well as in electrically excitable axonal membranes and in non-excitable epithelial cells, the opening of ≈ 1 pS conductance pathways, through which ≈ 10^4 elementary ionic charges pass in about 1 ms.

At an even deeper level, which is the atomic one, a purely physical noise appears, due to the thermal motion of the ions (The Johnson-Nyquist noise) and the discrete ionic charges.

As it was previously said, always fluctuations are an inherent consequence of the particulate nature or matter. The fact that, in living systems, they arise at several hierarchical levels suggests that a quantum-type behaviour (as, for instance, the "all-or-nothing" response of some excitable systems), with quantum-type parameters might characterize these systems.

If a random variable (the conductance G of a membrane area) is the sum of N independent and identical random variables (the conductances g of each channel), then the mean value G and the variance σ^2_G of G, are N times greater than the corresponding values of g and its variances.

$$G = N \cdot g \quad \text{and} \quad \sigma^2_G = N\,\sigma^2_g \qquad (VII.1)$$

The relative fluctuation,defined as the ratio between the standard deviation and the mean, is:

$$\sigma G/G = \sqrt{\sigma^2 G}/G = 1/\sqrt{N}$$

The fact that the relative fluctuations is proportional with $N^{-1/2}$ explains why fluctuations analysis has been much more successful with membrane structures than with reactions in homogeneous solutions.

All types of physico-chemical analyses give macroscopic

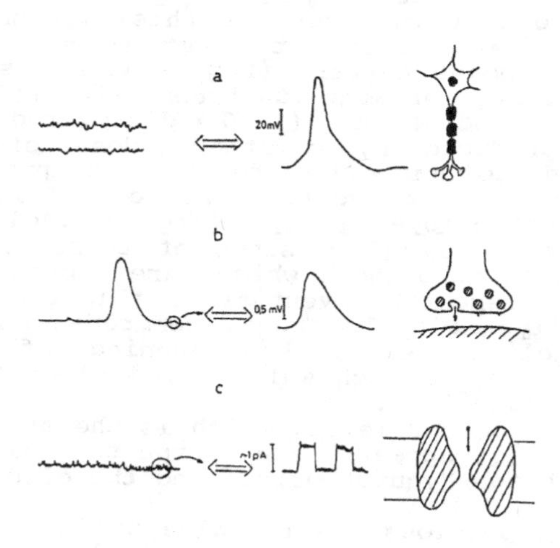

Figure VII.4. Schematic view of the individual electric events and their associated structures, at three levels: the cellular action potential (a); the miniature post-synaptic potential elicited from the release of a quantum of chemical transmitter (b), and the opening of a

chemically-operated ionic channel in the post-synaptic membrane (c). In the left part, there are the electric records from the corresponding higher level, in which those events appear as noise: the EEG (a), the record from microelectrodes placed in the postsynatpic area (b), and high-gain current recording in a voltage clamped membrane (from Margineanu and Vais, 1988).

membrane characteristics which result from spatial summation and temporal averaging of the above discussed single microscopic contributions, over the entire transporter population and the duration of the measurement. Indeed, the relation (VII.1) between the macroscopic conductance of a membrane area and the mean individual conductance of the channels within it, clearly reveals both the spatial summation over the N individual channels, and the temporal mean of the channel conductance. Thus, the relation is expected to hold only if the time spent for observing the transport phenomenon is sufficiently long as compared with the dwelling time in the open state of each channel. In other words, if one could obtain a quasi-instantaneous picture of a membrane area, its conductance should be correlated only with those channels open at that particular moment.

The integration performed by the whole membrane is a spatio-temporal one but the problem is how many channels must be present in a membrane patch in order to obtain a smooth behaviour. In a remarkable computer experiment, Clay and DeFelice (1983) have shown that the characteristic macroscopic behaviour of the axonal membrane only becomes apparent when several hundred (simulated) channels are averaged together. From Fig. VII.5, it appears that the action potential for 1 μm^2 patch area, containing 300 sodium channels and 60 potassium channels, is roughly similar in shape to the macroscopic action potential, while for a 0.04 μm^2 patch, the effects of single channel openings and closings are fully apparent. The authors concluded that on a surface of only 1 μm^2, the spatial summation smoothes the electrical behaviour of the nerve membrane and appear as a macroscopic response.

In connection with the problem of statistical averaging within neural systems, it should be observed that the integration of several organizational levels allows to combine a very high sensitivity with a stable functioning. There are several classical examples in sensory physiology showing the ability of specialized neurons to respond to one or a few light quanta or odorant molecules and recently in rat olfactory neurons the initiation of action potentials was observed by the opening of a single channel (Lynch and Barry, 1989). But, on the other hand, the successive hierarchical integration shields the whole neural system from the restless molecular motions.

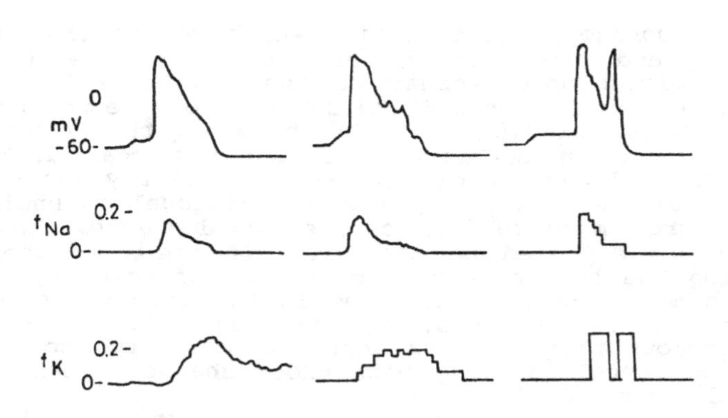

Figure VII.5. Simulated action potentials (upon row), based on the classical Hodgkin-Huxley model of the nerve impulse, corresponding to membrane areas of 1 μm² (left), 0.2 μm² (middle) and 0.04 μm² (right). The fractions of open sodium (fNa) and potassium (fK) channels are also given in each case.The effect of averaging to produce smoother variations is obvious for larger areas and is better manifested for sodium channels, which are only 60 μm^{-2} (redrawn from Clay and DeFelice, 1983).

References

Alkon, D.L. (1987) Memory Traces in the Brain. Cambridge University Press, London.
Alkon, D.L. (1989) Scientif. Amer. 261, 42-50.
Barchi, R.L. (1983) J. Neurochem. 40, 1377-1385.
Changeux, J.P. and Dunchin, A. (1976) Nature 264, 705-712.
Clay, J.R. and De Felice, J.L. (1983) Biophys. J. 42, 151-157.
Colynhoun, D. and Hawkes, A.G. (1983) Ch. 9 in: Single Channel Recording (B. Sackmann and E. Neher, Eds.),

Plenum Press, New York.
Drews, G. and Ruck, M. (1988) Biophys. J. 54, 383-391.
Elmer, L.E., O'Brien, B., Nutter, T.J. and Angelichs, K.J. (1985) Biochemistry 24, 8128-8137?
Hamill, O.P., Marty, A., Neher, E., Sackmann, B. and Sigworth, F.J. (1981) Pflƒgers Arch. 391, 85-100.
Hille, B. (1984) Ionic channels of Excitable Membranes (Ch. 3) Sinauer, Sunderland Mass.
Katz, B. and Miledi, R. (1970) Nature 226, 962-963.
Lester, H.A. (1988) Science 241, 1057-1063.
Llinks, R.R. (1988) Science 242, 1654-1664.
Lynch, ?
Margineanu, D.G., Katona, E. and Popa, J. (1981) Biochim. Biophys. Acta 649, 581-586.
Margineanu, D.G. and Vuis, H. (1988) Seminars in Biophysics 5, 131-142.
Neher, E. and Sackmann, B. (1976) Nature 260, 799-802.
Noda, M., Ikeda, T., Kayano, T., Suzuki, H., Takeshima, H., Kurasaki, M., Takahashi, H. and Numa, S. (1986) Nature 320, 188-192.
Numa, S. (1986) Biochem. Soc. Symp. 52, 119-143.
Takumi, T., Ohkubo, H. and Nakanishi, S. (1988) Science 242, 1042-1045.
Tokoshima, C. and Unwin, N. (1988) Nature 336, 247-250.

INDEX OF NAMES

INDEX OF SUBJECTS